GLOBE BIOLOGY

REVISED EDITION

LABORATORY PROGRAM
Annotated Teacher's Edition

LEONARD BERNSTEIN

GLOBE FEARON
Pearson Learning Group

Globe Biology Program
Globe Biology, Student Edition
Globe Biology, Annotated Teacher's Edition
Globe Biology, Laboratory Program, Student Edition
Globe Biology, Laboratory Program, Annotated Teacher's Edition
Globe Biology, Teacher's Resource Book
Globe Biology, Color Transparencies
Globe Biology, Foreign Language Supplement (Spanish)

Copyright © 1999 by Pearson Education, Inc., publishing as Globe Fearon®, an imprint of Pearson Learning Group, 299 Jefferson Road, Parsippany, NJ 07054. All rights reserved. No part of this book may be reproduced or transmitted in any form or by any means, electronic or mechanical, including photocopying, recording, or by any information storage and retrieval system, without permission in writing from the publisher. For information regarding permission(s), write to Rights and Permissions Department.

ISBN 0-8359-5745-4 (Laboratory Program, Student Edition)
ISBN 0-8359-5744-6 (Laboratory Program, Annotated Teacher's Edition)
Printed in the United States of America
4 5 6 7 8 9   06 05 04 03 02

1-800-321-3106
www.pearsonlearning.com

Table of Contents

Overview of the Globe Biology Laboratory Program	T-v
Laboratory Program Correlation to Skills Worksheets	T-vi
Guidelines for Safety	T-vii
Care of Live Animals	T-viii
Reading Skills and Laboratory Procedures	T-ix
Materials List	T-x
Suppliers Names and Addresses	T-xv
Preparation of Solutions	T-xvi

Overview of the Globe Biology Laboratory Program

The laboratory investigations in the **Globe Biology Laboratory Program** are designed to provide students with hands-on experiences in biology. The program gives students opportunities to make decisions, to use some of the tools associated with biology, and to implement scientific method.

Each of the 70 laboratory investigations is correlated to a concept or a process presented in the student text. The laboratory program consists of several types of laboratories. The variety of types of laboratories provides students with many of the experiences that biologists encounter as they work either in a laboratory or in the field.

Types of Laboratories

Among the types of laboratories presented are the following:

1. **Decision-Making Laboratories**

    Example: *Laboratory 3-1 How are the properties of mixtures and compounds different?*

    In the decision–making laboratories, students are given the opportunity to think through a problem and actually make a decision to solve the problem.

2. **Dissection Laboratories**

    Example: *Laboratory 13-1 What are the external and internal structures of an earthworm?*

    Dissection laboratories are presented for those teachers who would like to provide dissection experience. These laboratories are optional and do not need to be completed if teachers or students oppose dissection for moral, social, or religious reasons.

3. **Hands–On Experience Laboratories**

    Example: *Laboratory 21-1 How does saliva help in the chemical digestion of food?*

    In hands-on experience laboratories, students are afforded the opportunity to develop skills in manipulating laboratory equipment and to carry out procedures that lead to a specific outcome or conclusion.

4. **Modeling Exercises**

    Example: *Laboratory 20-1 What kinds of movements are possible for human body joints?*

    In these laboratory investigations, students construct models to help make certain biological concepts less abstract.

5. **Simulation Laboratories**

    Example: *Laboratory 22-2 Which blood types can be mixed safely?*

    In these laboratory investigations, students complete simulation–type activities, such as examining a given population and counting the members of the populations, or simulating inheritance patterns.

6. **Dry Laboratories**

    Example: *Laboratory 24-3 How can your senses be fooled?*

    In dry laboratories, students work with paper, pencil, diagrams, charts, and so on. They do not work with laboratory equipment, but instead work with concepts and ideas.

Format

Many of the laboratories in the **Globe Biology Laboratory Program** are designed to be completed in one class period. However, some laboratories require additional observation time during one or several days.

The format for each laboratory is outlined as follows:

- **Title:** Each heading is titled with a question, identifying the problem to be solved by students as they complete the laboratory.

- **Background Information:** Additional information that may be helpful to the students as they do the laboratory is provided.

- **Skills:** A list of the major skills that students will practice as they complete each laboratory is identified.

- **Objectives:** The objectives provide students with clearly stated goals. The objectives clearly define what the students will be doing and are expected to learn as they complete each laboratory.

- **Prelab Preparation:** This section of each laboratory provides students will additional information, suggestions, and activities to perform *before* beginning each laboratory.

- **Materials:** Each laboratory has a list of materials and equipment necessary to complete the laboratory. Quantities are provided when a specific number of materials or amount of material is important.

- **Procedure:** An easy–to–follow, step–by–step set of directions provides details to students and leads students through each laboratory. Caution statements are printed in boldfaced type and safety symbols are used throughout the laboratories to alert students to safety concerns.

- **Observations and Data:** In maintaining scientific method, observations and collected data are recorded in tables and charts.

- **Analysis and Conclusions:** Students are asked to analyze their data and observations to draw conclusions. The questions in this section relate directly to the title question and objectives.

- **Extension:** Each laboratory concludes with an extension, providing students with additional activities or laboratories. The extensions may be used as alternative laboratories or activities, supplemental exercises, or enrichment exercises.

Teacher Annotations

The Annotated Teacher's Edition provides annotations printed in boldfaced type on the student pages. Answers, teaching suggestions, and additional directions or safety guidelines are provided on the pages corresponding to the student pages.

Laboratory Program Correlation

Activity Skills Worksheets and Laboratory Skills Worksheets are provided in the **Globe Biology Teacher's Resource Book.** These worksheets provide students with an opportunity to develop and reinforce a variety of skills that can make their work in the biology laboratory more meaningful. Suggestions for where to implement each Activity Skills Worksheet and Laboratory Skills Worksheet into your laboratory program are listed below.

Activity Skills Worksheets

Worksheet Letter and Title	Page Numbers	Where to Use
A. Safety in the Biology Laboratory	A-3, A-4	Laboratory 1-1
B. Following Directions in an Activity	A-5, A-6	Laboratory 1-1
C. Writing a Hypothesis	A-7, A-8	Laboratory 3-2
D. Organizing and Analyzing Data	A-9, A-10	Laboratory 1-1
E. Making Observations and Inferences	A-11, A-12	Laboratory 3-2
F. Making and Labeling Diagrams	A-13, A-14	Laboratory 4-2
G. Making and Analyzing Graphs	A-15, A-16	Laboratory 32-1

Laboratory Skills Worksheets

Worksheet Number and Title	Page Numbers	Where to Use
1. Safety in the Science Laboratory	E/S 109-110	Use before Laboratory 1-1
2. Laboratory Equipment and Uses	E/S 111-112	Use before Laboratory 1-2
3. SI Measurement	E/S 113-114	Use before Laboratory 1-2
4. Measuring with a Compound Microscope	E/S 115-116	Use before Laboratory 1-1
5. Using Celsius Thermometers	E/S 117-118	Use before Laboratory 1-1
6. Caring for Live Animals	E/S 119-120	Use before Laboratory 13-2
7. Using a Bunsen Burner	E/S 121-122	Use before Laboratory 3-1
8. Designing an Experiment	E/S 123-124	Use before Laboratory 1-1
9. Writing a Laboratory Report	E/S 125-126	Use after Laboratory 1-1
10. Dissection Techniques	E/S 127-128	Use before Laboratory 13-1

Guidelines For Safety

Biology Laboratories are interesting and exciting learning experiences for students. The "hands-on" approach provided by laboratories reinforces what students learn in their text. As with all science laboratories, however, safety rules must be observed.

Safety is an integral element in the study of biology. To insure that adequate safety guidelines are followed, each state has general regulations for classroom safety that should be enforced. If you are not aware of these regulations, contact your state Department of Education. By practicing laboratory procedures in a safe manner, you will be providing students with a good model to follow.

Safety Equipment

At the beginnig of each term, make sure that the exits are clearly marked and easily accessible. The following equipment should be available in the biology laboratory and be in good working order:

- fire blanket
- eyewash station
- sand buckets
- complete first aid kit
- dry chemical fire extinguisher

Before students begin laboratory work, make them aware of the location and proper use of all safety equipment, first aid procedures, proper storage and disposal procedures for equipment and chemicals, and procedures in case of fire or other laboratory emergencies.

Also, provide each student with the following safety equipment:

- safety goggles
- heat-resistant gloves
- lab apron
- tongs or test tube holders

Safety Guidelines

Have students read through each laboratory investigation before beginning, so they will be aware of potentially dangerous situations. Review proper procedures and meanings of all safety symbols included in the laboratory investigations. No eating, drinking, or misconduct should be tolerated in the laboratory.

Proper Disposal Direct students to clean up work areas at the end of each activity; dispose of all laboratory materials.

Clothing Protection Have students wear lab aprons to protect clothing from burns or stains.

Glassware Safety Caution should be used when handling glassware; check all glassware for chips or cracks; dispose of broken glassware; be sure students do not force glass tubing into rubber stoppers; students should clean all glassware and air dry it.

Heating Safety Direct students to use caution when handling hot objects; follow proper procedures when lighting a Bunsen burner; caution students never to reach across an open flame; turn off heat sources when not in use; caution students who are heating chemicals in a test tube to point the test tube away from others.

Poison Caution students not to mix any chemicals without your instructions; store and dispose of poisons properly; be sure students inform you if any chemical gets in their eyes, touches their skin, or is swallowed.

Fire Safety Direct students to tie back loose hair and restrain loose clothing; make sure that students know the location and proper use of fire extinguishers and fire blankets.

Plant Safety Caution should be used when students collect and handle plants; unfamiliar plants should never be eaten even if they are classified as edible.

Electrical Safety Inspect all electrical equipment for loose plugs or worn cords; keep work areas dry; be sure students do not overload electric circuits.

Explosion Caution students not to mix chemicals together unless instructed to do so; students should use water-baths to heat solids; be sure open flames are never used when working with flammable liquids.

Sharp Objects Students should be cautioned when handling any sharp tool, such as scissors, scalpels, knives, or other cutting instruments; first-aid procedures for cuts and puncture wounds should be reviewed with students.

Caustic Substances Review first aid for burns; instruct students to wash skin immediately if they spill acids or bases on themselves; direct students to wear laboratory aprons at all times; remind students never to pour water into an acid or a base.

Animal Safety Direct students to use care when handling live animals; notify the school nurse immediately if a student is bitten by an animal; students should never bring live animals into class unless the animals have been purchased from a reputable supplier.

Eye Safety Caution students to wear laboratory goggles at all times when using open flames and chemicals; review with students the procedure for using the emergency eyewash system.

Cleanup Remind students to wash their hands after every laboratory.

Proper Care and Use of Live Animals

An important characteristic of all laboratory programs in biology is the study of live animals. Live animals, however, must be cared for properly, as well as receive humane treatment. The proper handling of live animals will protect both the animals and your students. Also, if students learn how to care for animals in the laboratory, they may develop an appreciation for animals living outside the academic environment.

1. Do not perform surgical procedures on live animals in the laboratory.

2. Do not expose animals to toxic chemicals, harmful drugs, pathogens, or radiation.

3. Consider the safety of animals during all laboratory procedures.

4. Research and provide proper environments for different kinds of animals.

5. Handle animals gently and only when necessary. Special handling is necessary for pregnant or feeding animals, or animals that become easily frightened.

6. Wear gloves when handling rodents to prevent injury from bites.

7. Give newly-acquired animals several days to acclimate themselves to their new home before handling them.

8. Place fish and other aquatic organisms shipped in water-filled containers in an aquarium that is the same temperature as the container they were shipped in.

9. Do not remove wild reptiles, birds, or mammals from their natural habitats. These animals are very difficult to care for in a laboratory.

10. Invertebrates and amphibians may be captured for laboratory use if not prohibited by local, state, or federal law. However, always return these animals to their natural habitats during the appropriate season.

11. Do not place cages or tanks near open windows, radiators, or air conditioners.

12. Place glass containers with animals in locations that do not receive direct sunlight to prevent overheating.

13. Provide cold-blooded animals, such as frogs and snakes, with enough heat during the late fall, winter, and early spring.

14. Provide animals with enough space and materials for bedding, nesting, and gnawing.

15. Provide small mammals, such as mice, hamsters, and gerbils, with exercise wheels.

16. Provide for the needs of animals over weekends and during school vacations. Keep a record of who is feeding the animals; when the animals are being fed; and how much food the animals are getting at each feeding.

17. Provide animals with clean containers and an adequate supply of clean water.

18. Provide particular species of animals with the proper kind and amount of food.

19. Do not overfeed fish. Fish can be fed once or twice a week.

20. Purchase animals from a respectable pet store or biological supply house.

Reading Skills and Laboratory Procedures

Many biology students have difficulty reading and understanding laboratory procedures. These students often require more time, guidance, and positive reinforcement to ensure their success. An effective and simple method to use to help discover students who may have reading comprehension problems is to have all students first read the entire laboratory. On a separate sheet of paper, have students write all unfamiliar words and procedural steps that are unclear to them. Review these unfamiliar words and procedural steps to ensure students success in completing the laboratory. You may also find the following suggestions helpful in addressing the needs of students with reading difficulties.

- Direct students to follow along as you read aloud the title, safety guidelines, and procedure for each laboratory investigation.

- Ask students to point out the safety guidelines for each laboratory investigation.

- Explain to students how each laboratory investigation relates to material in the text.

- Ask students to define new terms in each laboratory investigation.

- Illustrate the proper and safe use of equipment and supplies. As you read through the procedure, point out what each piece of equipment looks like and how it is used.

- Have students keep a record of new words and terms used in laboratory investigations. Index cards can then be compiled in alphabetical order and used by students as a study aid. Keep the index cards in a resource area of the laboratory.

- Encourage students to answer questions in complete sentences.

Equipment List

The quantities are for a class of 30 students.

Equipment	Quantity	Used in Laboratory
ammonia	6 L	8-1
antacids	1 package	28-1
baby food jars	15	8-1, 11-1, 28-2
balances	15	3-1
bar magnets	30	3-1
beakers (250 mL)	45	1-2, 10-2, 23-1, 23-2
beans, dried	2 bags	29-1
begonia plant	15	11-2, 11-3
Benedict's solution	100 mL	3-3, 21-1, 26-1
binoculars	15	18-2
Biuret solution	100 mL	3-3
black paper	1 package	13-2
bleach	2 L	8-1, 12-2
blue food coloring	1 L	22-2
bottles, plastic (2 L)	15	23-1
bread	1 loaf	8-1
bromthymol blue solution	2 L	2-1, 8-2, 23-2
brown wrapping paper	1 roll	3-3
Bunsen burner	15	3-1, 3-2, 3-3, 21-1, 26-1
cards, index	1 package	1-2, 24-2, 32-2
cigarette, cigar, and pipe tobacco	assorted	28-2
clay triangles	15	3-2
clear plastic cups	30	11-2
coins	100	27-2, 30-3, 31-2
compound microscopes	15	1-1, 2-1, 4-1, 4-2, 7-1, 7-2, 8-2, 10-2, 12-1, 12-2, 22-1, 27-1, 30-2, 35-2, 37-2
Congo red solution	1 L	28-1
conifer twigs	30	32-1
construction paper	1 package	4-2, 31-2
corn syrup	2 L	26-1
cotton balls	1 package	7-1, 24-2
cotton swabs	1 package	6-1, 8-1, 13-2, 27-2
cover slips	60	1-1, 2-1, 4-1, 7-1, 8-2, 12-1, 12-2, 27-1, 35-2
crackers	1 box	8-1
crucibles	15	3-2
culture of live amoebas	1	7-1
culture of live *Chorella*	1	35-2
culture of live *Didinium*	1	35-2
culture of live paramecia	1	7-1, 35-2

Equipment	Quantity	Used in Laboratory
depression slides	30	12-1, 35-2
detergent	1 box	37-1
dilute ammonia solution	2 L	23-2
dissecting microscope	6	1-1
dissecting needles	15	1-1, 13-1, 15-1, 15-2, 16-1, 17-1
dissecting pans	15	13-1, 13-2, 14-1, 15-1, 15-2, 16-1, 17-1
droppers	30	1-1, 3-3, 4-1, 7-1, 8-1, 8-2, 12-1, 12-2, 13-2, 14-1, 21-1, 22-2, 23-2, 26-1, 27-1, 28-1, 35-2
Elodea	15	4-1
earthworm	15	13-2
ethanol	2 L	28-2
feathers from different birds	assorted	18-1
field guides to birds	5	18-2
filter paper	1 box	3-1, 3-2, 3-3
flashlights	15	12-1, 13-2
food colorings, assorted	1 L	10-2, 23-1
food labels	assorted	26-2
forceps	15	4-1, 11-3, 13-1, 14-1, 16-1, 17-1
funnels	15	3-1, 3-3
glass tubing, 10 cm	3 m	23-1
glass tubing, 20 cm	3 m	23-1
glue	1 bottle	4-2
graduated cylinders (10 mL)	15	2-1
graduated cylinders (100 mL)	50	1-2, 2-1, 3-1 3-2, 3-3, 8-2, 23-1, 23-2, 27-1, 28-1
Grantia	15	12-2
graph paper	1 package	32-1, 37-2
hand lenses	15	1-1, 2-1, 3-2, 3-3, 6-1, 8-1, 9-1, 11-1, 11-3, 13-1, 14-1, 15-1, 15-2, 16-1, 18-1, 27-2, 28-2
household ammonia	1 L	37-1
household cleaners, 2	1 box	8-1
hydra culture	1	12-1
hydrochloric acid	1 L	28-1
iodine stain	1 bottle	4-1
iron filings	10 g	3-1
lead nitrate	1 g	3-1
lima beans	1 package	11-1, 28-2, 37-1
Lugol's solution	100 mL	3-3, 21-1, 26-1
marking pencils	15	2-1, 6-1, 8-1, 8-2, 11-1, 21-1, 22-2, 27-1, 27-2, 28-1, 28-2 37-1, 37-2
masking tape	1 roll	37-1

Equipment	Quantity	Used in Laboratory
matches	1 box	3-2
metersticks	15	24-2
metric rulers	15	9-1, 10-1, 10-2, 11-1, 24-2, 24-3, 28-2, 23-1, 33-1
microscope slides	15	1-1, 2-1, 4-1, 7-1, 8-2, 12-2, 27-1, 37-2
milk	1 L	27-1
moss plants	assorted	9-1
nutrient agar	2 L	6-1, 27-2
oatmeal	1 box	8-1
onions	15	11-2
oranges	15	8-1
paint thinner	2 L	3-2
paper clips	3 boxes	1-2
paper cups	3 packages	9-1, 22-2
paper towels	3 rolls	4-1, 11-1, 13-2, 19-1, 28-2, 37-1
Petri dishes	30	1-2, 2-1, 6-1, 27-2
petroleum jelly	1 jar	37-2
pins	30	13-1, 17-1, 24-2
pinto beans	1 bag	32-1
pipe cleaners	1 roll	4-2
plastic knives	15	8-1
plastic sandwich bags	15	37-1
plastic spoons	15	2-1, 3-2, 8-1
pond water	2 L	1-1
potatoes, white	2	4-1, 11-2
potting soil	1 bag	9-1
prepared slide of amoebas	15	7-1
prepared slide of blood showing showing sickle-cell anemia	15	30-2
prepared slide of blood showing sickle-cell trait	15	30-2
prepared slide of *Euglena*	15	7-2
prepared slide of herbaceous stems	15	10-2
prepared slide of human blood	15	1-1, 22-1
prepared slide of human skin cells	15	4-1
prepared slice of normal human blood	15	30-2
prepared slide of onion root tip	15	4-2
prepared slide of paramecia	15	7-1
prepared slide of *Spirogyra*	15	7-2
prepared slide of Volvox	15	7-2
prepared slide of woody stems	15	10-2

Equipment	Quantity	Used in Laboratory
preserved crayfish	15	15-1
preserved earthworm	15	13-1
preserved frog	15	17-1
preserved grasshopper	15	15-2
preserved insect	15	1-1
preserved perch	15	16-1
preserved sea star	15	14-1
probes	15	18-1
ptyalin solution (amylase)	1L	21-1
raw potatoes	15	4-1, 26-1
razor blades	15	4-1, 10-2, 11-2
red food coloring	1 L	22-2
rice	1 box	8-1
rings	15	3-2
ring stands	15	3-2
roses	15	11-3
rubber hammers	15	24-2
rubber tubing, 30 cm	10 m spool	23-1
salt, table	1200 g	3-1
sand	2 bags	3-1
scalpels	15	2-1, 11-3, 13-1
scissors	15	2-1, 4-2, 9-1, 13-1, 14-1, 16-1, 17-1, 19-1, 24-2, 31-1, 31-2, 33-1
shallow dishes	15	3-3
shallow pans	15	19-1
small beakers	15	3-1, 12-2
sodium bicarbonate solution	60 g	28-1
sodium iodide	6 g	3-1
soil	2 bags	19-1
Spongia	15	12-2
stalks of celery with leaves	15	10-2
starch solution	50	21-1
stirring rods	15	23-2, 28-1
stoppers, rubber 2 hole	15	23-1
straws	1 box	23-2
sugar solution	15 mL	8-2
sulfur	15	3-1
sulfuric acid	160 mL	37-1
tape, cellophane	several rolls	4-2, 30-3, 31-1, 37-2
test tube holders	15	3-1, 3-3, 21-1, 26-1
test tube racks	15	2-1, 3-2, 3-3, 8-2, 21-1, 26-1, 27-1, 28-1

Equipment	Quantity	Used in Laboratory
test tubes	75	2-1, 3-1, 3-2, 3-3, 8-2, 21-1, 26-1, 27-1, 28-1
tomato slices	2	4-1
tongs	15	3-2
toothpicks	1 box	11-2, 12-1, 12-2, 13-2
triple beam balances	15	1-2
tulips	5	11-3
tuning forks	15	24-2
twigs of pine, spruce, fir, and cedar	assorted	10-1
vinegar	3000 mL	11-1, 12-1, 13-2, 37-1
watch or clock with second hand	15	23-2, 28-3
waxed paper	1 box	26-1
yarn	1 skein	4-2
yeast suspension	15	8-2

Laboratory Suppliers and Addresses

American Biological Supply Company
288 East Green Street
Westminster, Maryland 21157
Phone: 1-800-299-2624
Fax: 1-410-876-3438
www.qis.net/~ambi

Carolina Biological Supply Co.
2700 York Road
Burlington, NC 27215
Phone: 1-800-334-5551
Fax: 1-800-222-7112
www.carosci.com

Fisher Scientific Co.
Stansi Educational Material Division
485 South Frontage Road
Burr Ridge, Illinois 60521
Phone: 1-800-955-1177
Fax: 1-800-926-1166
www.fisher1.com

Frey Scientific Company
905 Hickory Lane
Mansfield, Ohio 44905
Phone: 1-800-255-3739
Fax: 1-419-589-1522
www.freyscientific.com

Nasco Science
901 Janesville Avenue
P.O. Box 901
Ft. Atkinson, Wisconsin 53538-0901
Phone: 1-800-558-9595
Fax: 1-414-563-8296
www.nascofa.com

Sargent Welch Scientific Company
P.O. Box 5229
Buffalo Grove, Illinois 60089-5229
Phone: 1-800-727-4368
Fax: 1-800-676-2540
www.sargentwelch.com

Wards Scientific Establishment, Inc.
PO Box 92912
Rochester, NY 14692-9012
Phone: 1-800-962-2660
Fax: 1-800-635-8439
www.wardsci.com

Preparation of Solutions

Laboratory	Solution	Preparation
2-1, 8-2, 23-2	Bromthymol blue solution	Add 0.5 g bromthymol blue to 1 L distilled water.
3-3	"Unknown" solution	Combine 15 mL distilled water, 1 mL vegetable oil, 1 mL egg white, 1 g table salt, 1 mL corn syrup, and 1 g corn starch.
3-3, 21-1, 26-1	Lugol's solution	Add 1 g iodine crystals and 3.5 g potassium iodide to 300 mL distilled water.
3-3	Biuret solution	Combine 1L 10% potassium hydroxide solution and 25 mL 3% copper sulfate solution.
4-1	Iodine stain	Dissolve 1.5 g potassium iodide and 0.3 g iodine in 1 L water.
6-1, 27-2	Nutrient agar	Use agar and mix according to package instructions with distilled water. Pour into Petri dishes while warm.
21-1	Starch solution	Add 3 g corn starch to 500 mL cold distilled water. Stir thoroughly and heat to boiling.
28-1	Congo red solution	Add 3 g congo red dye to 100 mL distilled water.
28-1	Sodium bicarbonate solution	Add 1 mL sodium bicarbonate to 1 mL distilled water. Stir to dissolve.
28-1	Hydrochloric acid solution	Add 35 mL concentrated hydrochloric acid to 500 mL distilled water. **CAUTION: Do not add water to acid.**
28-1	Antacid solution	Dissolve one antacid tablet in 200 mL distilled water.
37-1	Detergent solution	Add 250 mL liquid detergent to 250 mL tap water.
37-1	Ammonia solution	Add one part ammonia to two parts tap water.

GLOBE BIOLOGY

REVISED EDITION

LABORATORY PROGRAM

LEONARD BERNSTEIN

Upper Saddle River, New Jersey
www.globefearon.com

Copyright © 1999 by Globe Fearon, Inc., One Lake Street, Upper Saddle River, New Jersey, 07458, www.globefearon.com. All rights reserved. No part of this book may be reproduced or transmitted in any form or by any means, electronic photographic, mechanical, or otherwise including photocopying, recording, or by any information storage and retrieval system, without permission in writing from the publisher.

ISBN 0-835-95745-4 (Laboratory Program, Student Edition)
ISBN 0-835-95744-6 (Laboratory Program, Annotated Teacher's Edition)

Printed in the United States of America.

2 3 4 5 6 7 8 9 10   04 03 02 01

Table of Contents

Chapter 1 The Science of Biology

Laboratory 1-1 How do you use different microscopes to examine different specimens? 1

Laboratory 1-2 How do you measure mass and volume? 5

Chapter 2 The Nature of Living Things

Laboratory 2-1 How can you determine if an unknown is living or nonliving? 9

Chapter 3 The Chemistry of Life

Laboratory 3-1 How are the properties of mixtures and compounds different? 11

Laboratory 3-2 How can you identify inorganic and organic compounds? 15

Laboratory 3-3 How can you identify the compounds found in living things? 19

Chapter 4 Cells

Laboratory 4-1 How do plant cells and animal cells differ in structure and function? 21

Laboratory 4-2 How do mitosis and meiosis compare? 25

Chapter 5 Classification

Laboratory 5-1 How can a dichotomous key be used to identify an organism? 27

Chapter 6 Viruses and Monerans

Laboratory 6-1 Where are bacteria found in your environment? 31

Chapter 7 Protists

Laboratory 7-1 How do different kinds of protozoans compare? 33

Laboratory 7-2 How do different kinds of algae compare? 35

Chapter 8 Fungi

Laboratory 8-1 How can the growth of molds be slowed or stopped? 37

Laboratory 8-2 How do yeast cells reproduce and carry on respiration? 41

Chapter 9 Spore Plants

Laboratory 9-1 How does the amount of available moisture and light affect the growth of mosses? 43

Chapter 10 Seed Plants

Laboratory 10-1 How can a dichotomous key be used to identify different kinds of conifers? 45

Laboratory 10-2 How are xylem and phloem arranged in woody and herbaceous stems? 47

Chapter 11 Plant Structure and Function

Laboratory 11-1 How do different conditions affect the rate of germination of lima bean seeds? 51

Laboratory 11-2 How do plants reproduce by vegetative propagation? 53

Laboratory 11-3 What are the parts of a perfect and an imperfect flower? 55

Chapter 12 Porifera and Cnidarians

Laboratory 12-1 How does a hydra respond to stimuli? 59

Laboratory 12-2 What do the support structures of sponges look like? 61

Chapter 13 Worms

Laboratory 13-1 What are the external and internal structures of an earthworm? 63

Laboratory 13-2 How does an earthworm respond to stimuli? 67

Chapter 14 Mollusks and Echinoderms

Laboratory 14-1 What are the external and internal structures of a sea star? 71

Chapter 15 Arthropods

Laboratory 15-1 How do the external features of a crayfish help to identify it as an arthropod? 75

Laboratory 15-2 How do the external features of a grasshopper help to identify it as an arthropod? 77

Chapter 16 Fishes

Laboratory 16-1 How do the external and internal structures of a perch adapt it for life in the water? 79

Chapter 17 Amphibians and Reptiles

Laboratory 17-1 How do the external and internal structures of a frog adapt it for life both in and out of the water? 83

Chapter 18 Birds

Laboratory 18-1 How do different bird feathers compare? 87

Laboratory 18-2 How are birds adapted to their environment? 89

Chapter 19 Mammals

Laboratory 19-1 What kinds of tracks do animals leave? 91

Laboratory 19-2 How are placental mammals classified? 95

Chapter 20 Support and Movement
Laboratory 20-1 What kinds of movement are possible for human body joints? 99

Chapter 21 Digestion
Laboratory 21-1 How does saliva help in the chemical digestion of food? 103

Chapter 22 Circulation
Laboratory 22-1 What kinds of cells are found in human blood? 105
Laboratory 22-2 Which blood types can be mixed safely? 107

Chapter 23 Respiration and Excretion
Laboratory 23-1 How much air do you normally exhale? 109
Laboratory 23-2 What effect does exercise have on the amount of carbon dioxide you exhale? 113

Chapter 24 Regulation
Laboratory 24-1 Which side of your brain is dominant? 115
Laboratory 24-2 How can you test your senses of touch and hearing? 117
Laboratory 24-3 How can your senses be fooled? 121

Chapter 25 Reproduction and Development
Laboratory 25-1 What changes take place during childbirth? 125

Chapter 26 Nutrition
Laboratory 26-1 Which foods contain sugar and starch? 129
Laboratory 26-2 What is the purpose of food additives? 133

Chapter 27 Diseases and Disorders
Laboratory 27-1 What conditions can prevent milk from spoiling? 137
Laboratory 27-2 What are some ways in which diseases can be spread? 139

Chapter 28 Drugs, Alcohol, and Tobacco
Laboratory 28-1 Which antacids are most effective in neutralizing acid? 141
Laboratory 28-2 How do alcohol and tobacco affect the germination of seeds? 143
Laboratory 28-3 What substances are produced when tobacco burns? 145

Chapter 29 Fundamentals of Genetics
Laboratory 29-1 How can a test cross help to determine genotype? 147
Laboratory 29-2 How can the results of monohybrid and dihybrid crosses be predicted? 151

Chapter 30 Modern Genetics
Laboratory 30-1 What percentage of students in your class have inherited some common genetic traits? 155
Laboratory 30-2 How do normal red blood cells and sickled cells compare? 157
Laboratory 30-3 How are sex-linked traits inherited? 161

Chapter 31 Applied Genetics
Laboratory 31-1 How can a karotype be used to identify human genetic disorders? 165
Laboratory 31-2 How can you make a model of controlled animal breeding? 171

Chapter 32 Theories of Evolution
Laboratory 32-1 How do individuals of the same species vary? 173
Laboratory 32-2 What is natural selection? 177

Chapter 33 Evidence for Evolution
Laboratory 33-1 What is the half-life of a radioactive element? 179
Laboratory 33-2 How do living things provide evidence for evolution? 181

Chapter 34 Human Change Through Time
Laboratory 34-1 How are humans similar to and different from other primates? 185

Chapter 35 Ecosystems
Laboratory 35-1 What are food chains? 187
Laboratory 35-2 What is the relationship between a predator and its prey? 189

Chapter 36 Biomes
Laboratory 36-1 How are organisms adapted to survive in different biomes? 191

Chapter 37 Conservation
Laboratory 37-1 How do pollutants in water affect the germination of seeds? 195
Laboratory 37-2 What kinds of pollutants are in the air? 199

Safety in the Science Laboratory

Science is a learning process. Unlike other fields of study, science often allows you an opportunity to "learn by doing." Both in the laboratory and in the field you can carry out hands-on investigations. In the science classroom, you make things happen.

Working in the science laboratory can be an interesting and meaningful experience. However, in the laboratory, you may work with materials and equipment that can be dangerous if not handled properly. For this reason, safety must always be your first priority.

You can avoid accidents in the science laboratory by following these simple guidelines:

- Always handle materials carefully.
- Do not perform any laboratory investigation without direction from your teacher.
- Never work in the science laboratory alone.
- Always read all directions in an investigation before beginning the investigation.

You will see safety symbols in some of the laboratory investigations. Before beginning each investigation, always read the investigation and note any safety symbols and caution statements. You should know what each symbol means and always follow the guidelines that apply to each symbol.

Safety Symbols

1. Disposal
   - Keep your work area clean at all times.
   - Dispose of all materials properly. Follow the instructions for disposal given by your teacher.

2. Clothing Protection
   - Wear your laboratory apron to protect your clothing from stains or burns.

3. Humane Treatment
   - Always treat all animals in as humane a way as possible.
   - Make sure all animals receive the appropriate care necessary for their survival.

4. Glassware Safety
   - Handle glassware carefully.
   - Check all glassware for chips or cracks before using it.
   - Do not try to clean up broken glassware. Notify your teacher if you break a piece of glassware.
   - Air-dry all glassware. Do not use paper towels to dry glassware.
   - Never force glass tubing into the hole of a rubber stopper.

5. Heating Safety
   - Be careful when handling hot objects.
   - Turn off the Bunsen burner or other heat source when you are not using it.
   - Use the proper procedures when lighting a Bunsen burner.
   - When you heat chemicals in a test tube, always point the test tube away from people.
   - Use a ringstand between a glass flask or beaker and the flame of a Bunsen burner.

6. Poison
   - Never mix chemicals without directions from your teacher.
   - Use all poisonous chemicals with extreme caution.
   - Inform your teacher immediately if you spill chemicals or get any chemicals in your eyes or on your skin.
   - Do not eat or drink in the laboratory.

7. Fire Safety
   - Confine loose clothing and tie back long hair when working near an open flame.
   - Be sure you know the location of fire extinguishers and fire blankets in the laboratory.
   - Never reach across an open flame.

8. Plant Safety
   - Never eat any part of a plant that you cannot identify as edible.
   - Some plants, such as poison ivy, are harmful if they are touched or eaten. Use caution when handling or collecting plants. Always use a reliable field guide to plants.

9. Electrical Safety
   - Check all pieces of electrical equipment for loose plugs or worn cords before using them.
   - Be sure that electrical cords are not placed where people can trip over them or where cords can tip over laboratory equipment.
   - Do not use electrical equipment with wet hands or near water.
   - Never overload an electrical circuit.

10. Explosion
    - Use a water-bath to heat solids.
    - Do not use an open flame if you are working with a flammable liquid.
    - Do not mix chemicals together unless instructed to do so by your teacher.

11. Sharp Objects
    - Be careful when using scissors, scalpels, knives, or other cutting instruments.
    - Always dissect specimens in a dissecting pan. Never dissect a specimen while holding it in your hand.
    - Always cut in the direction away from your body.

12. Caustic Substances
    - Use extreme care when working with acids and bases. Both acids and bases can cause burns. If you spill an acid or a base on your skin, flush your skin with plenty of water. Notify your teacher immediately.
    - Never mix acids and bases unless you are instructed to do so by your teacher.
    - Never pour water into an acid or a base. Always pour an acid or a base into water.

13. Animal Safety
    - Be careful when handling live animals. Some animals can injure you or spread disease.
    - Do not bring live animals into class that have not been purchased at a reputable pet store.

14. Clean Up
    - Always wash your hands after an activity in which you handle chemicals, animals, or plants.

15. Eye Safety
    - Wear your laboratory goggles, especially when working with open flames and chemicals.
    - If chemicals get into your eyes, flush them with plenty of water. Notify your teacher immediately.
    - Be sure you know how to use the emergency eye-wash system in the laboratory.

Name _____ Class _____ Date _____

Chapter 1   The Science of Biology

Laboratory 1-1   How do you use different microscopes to examine different specimens?

Background Information

Most high school biology laboratories have three kinds of microscopes: hand lenses, dissecting microscopes, and compound microscopes. A hand lens is a simple microscope. You can examine details of a large specimen with a hand lens. A dissecting microscope has two eyepieces. You look into a dissecting microscope the same way you look into binoculars. With a dissecting microscope, you can see a three-dimensional image of a specimen. The compound microscope is the most commonly used microscope. You can use it to look at small specimens mounted on microscope slides.

Skills: observing, classifying, organizing, and recording data

Objectives

In this laboratory, you will
- learn how to use the three kinds of microscopes.
- prepare a wet-mount slide.
- select the best microscope to observe four different specimens.

Prelab Preparation

1. Review Section 1-4 Tools of the Biologist.

2. Define the term "field of view."

   <u>the circle of light seen through a microscope</u>

3. Review the parts of a compound microscope.

Materials

hand lens	dissecting microscope	compound microscope
microscope slide	cover slip	dissecting needle
dropper	pond water	preserved insect
prepared slide of human blood		

Procedure

Part A: Hand Lens

1. Hold a hand lens between an object, such as a pencil, and your eye.

2. Move the hand lens closer to or farther away from the object until the object comes into focus.

Part B: Dissecting Microscope

1. Raise the tube of a dissecting microscope by turning the adjustment knobs toward you.

2. Place a specimen on the stage of the dissecting microscope.

3. Slowly lower the microscope tube by turning the adjustment knobs away from you. Be sure that the lenses do not touch the specimen.

Biology   Copyright ©   by Globe Book Company

4. Adjust the eyepieces so that they are the correct distance apart for your eyes. Look through the eyepieces and identify the field of view.

5. Using the adjustment knobs, slowly raise the microscope tube until the specimen comes into focus.

Part C: Compound Microscope

1. Place a prepared slide on the stage of a compound microscope. Use the clips to hold the slide in place.

2. Use the course adjustment knob to bring the specimen into view.

3. Adjust the amount of light passing through the stage of the microscope.

4. Use the fine adjustment knob to bring the specimen into focus. Be sure to keep both eyes open when looking through the eyepiece.

Part D: Selecting a Microscope

1. Select the best microscope to observe the skin on the back of your hand. Draw what you see in Plate 1.

2. Select the best microscope to observe the structures of a preserved insect. Examine the insect carefully with the microscope you have chosen.

3. Select the best microscope to examine a prepared slide of human blood. Draw what you see in Plate 2.

4. Using a dropper, place a drop of pond water in the center of a clean, dry slide.

5. Place one edge of a cover slip on the slide.

6. Using a dissecting needle, slowly lower the cover slip onto the drop of pond water.

7. Select the best microscope to examine the slide of pond water. Draw what you see in Plate 3.

8. Complete Data Table 1.

Observations and Data

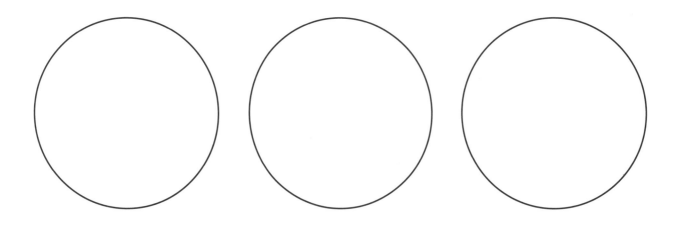

Plate 1: Human Skin   Plate 2: Human Blood   Plate 3: Pond Water

Name_____ Class_____ Date_____
Chapter 1   The Science of Biology   Laboratory 1-1

Data Table 1:   Selecting a Microscope

Specimen	Microscope
Skin	hand lens
Insect	hand lens/dissecting microscope
Blood	compound microscope
Pond water	compound microscope

1. Describe the appearance of the skin on the back of your hand.
   Students may describe seeing wrinkles, hair, and pores.

2. Describe two of the structures of the insect you examined.
   Answers will vary.

3. About how many blood cells did you see in the microscope's field of view?
   Answers will vary.

4. Describe the appearance of the blood cells you observed.
   Students may describe seeing red donut-shaped cells and larger, white or clear cells.

5. Did you see any living things in your sample of pond water? Describe them.
   Answers will vary.

Biology

Analysis and Conclusions

1. Explain why you selected the microscope you did to observe each of the following specimens:

    a. skin — **Specimens a and b were not transparent enough to be seen with the compound microscope. Specimens c and d were too small to be seen with the dissecting microscope.**

    b. insect

    c. blood

    d. pond water

2. Did you have any problems using different microscopes and making your observations? If so, describe them.
   Answers will vary.

Extension

Design an experiment to measure the length of a preserved insect using a clear metric ruler and a dissecting microscope. Describe your experiment in the space provided.

Name _____ Class _____ Date _____

Chapter 1 — The Science of Biology

Laboratory 1-2   How do you measure mass and volume?

Background Information

Mass is a measurement of the amount of matter in an object. You can measure the mass of an object using a triple beam balance. The mass of an object is measured in grams or kilograms. Volume is the amount of space occupied by a solid, liquid, or gas. You can measure the volume of a liquid using a graduated cylinder. Liquid volume is usually measured in milliliters or liters.

Skills: predicting, measuring, inferring

Objectives

In this laboratory, you will
- measure the mass of different objects.
- measure the mass of a given volume of water.
- measure the volume of a liquid.

Prelab Preparation

1. Review Section 1-3 Scientific Measurements.

2. Define the term "meniscus."

    the curved surface of a column of liquid in a container

Materials

triple beam balance	250-mL beaker	graduated cylinder
Petri dish	paper clip	index card
water		

Procedure

Part A: Using a Triple Beam Balance

1. Look at the triple beam balance shown in Figure 1. Identify the riders, beams, and pointer.

2. Be sure that all the riders and the pointer are at zero.

3. Predict the mass of a Petri dish. Record your prediction in Data Table 1.

4. Place the Petri dish on the pan of the balance.

5. Move the rider on the back beam one notch at a time until the pointer drops below the zero mark. Then move the rider back one notch.

6. Slowly slide the rider along the front beam until the pointer stops at the zero mark.

7. Calculate the actual mass of the Petri dish by adding the masses shown on the two beams. Record the actual mass in Data Table 1.

8. Remove the Petri dish from the pan.

Biology   Copyright ©   by Globe Book Company   5

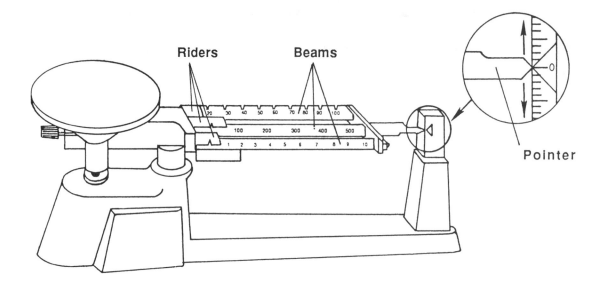

Figure 1: Triple Beam Balance

9. Predict the mass of a paper clip and of an index card. Record your predictions in Data Table 1.

10. Use the triple beam balance to find the actual mass of the paper clip and of the index card. Record the actual masses in Data Table 1.

11. Predict the mass of a 250-mL beaker. Record your prediction in Data Table 1.

12. Find the actual mass of the beaker. Record the actual mass in Data Table 1.

13. Predict the mass of 50 mL of water. Record your prediction in Data Table 1.

14. Pour water into the beaker up to the 50-mL mark.

15. Use the balance to find the mass of the beaker and water. Record this mass in Data Table 1.

16. To find the mass of the water, subtract the mass of the empty beaker from the mass of the beaker and water. Record the mass of the water in Data Table 1.

Part B: Using a Graduated Cylinder

1. Look at the graduated cylinder shown in Figure 2. Identify the meniscus.

2. Pour water into a 250-mL beaker until the beaker is about one-third full.

3. Carefully pour the water from the beaker into a graduated cylinder.

4. Find the volume of the water by reading the marking on the cylinder that is in line with the bottom of the meniscus. Record this volume in Data Table 2.

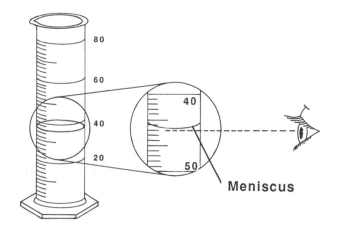

Figure 2: Graduated Cylinder

Name _____ Class _____ Date _____

Chapter 1   The Science of Biology   Laboratory 1-2

5. Pour the water from the cylinder back into the beaker. Add water to the beaker until it is about two-thirds full.

6. Use the graduated cylinder to find the volume of water in the beaker. Record this volume in Data Table 2.

Observations and Data

Data Table 1:   Mass

Object	Mass (grams)	
	Predicted	Actual
Petri dish	Answers will vary.	
Paper clip		
Beaker (empty)		
Beaker and water		
Water		

Data Table 2:   Volume

Beaker	Volume (mL)
One-third full	about 83 mL
Two-thirds full	about 167 mL

Analysis and Conclusions

1. What mass does each number on the first beam of the balance represent? on the middle beam? on the third beam?

   1 gram, 100 grams, 10 grams

2. Why do you think the tool used to measure mass is called a "balance"?

   because the mass of the object being measured is balanced by the masses on the beams

3. Suppose you want to add 6 mL of water to a test tube. Which of the following graduated cylinders would you use to measure this volume of water: a 10-mL cylinder, a 50-mL cylinder, or a 100-mL cylinder? Explain your answer.

   the 10-mL graduated cylinder, because the finer graduations would give a more accurate measurement

Biology   Copyright ©   by Globe Book Company

Extension

Design an experiment to find the volume of a solid object. Use a 10-mL graduated cylinder, water, and an object such as a nut or bolt. Record your observations in a data table. (Hint: A solid object displaces a volume of water equal to the volume of the object.) Describe your experiment in the space provided.

Name_____ Class_____ Date_____

Chapter 2   The Nature of Living Things

Laboratory 2-1   How can you determine if an unknown is living or nonliving?

Background Information

"Living" is often defined as the ability to carry out certain chemical reactions called life processes. Two life processes are respiration and growth. During the process of respiration, oxygen combines with food to release energy and carbon dioxide. You can test for the presence of carbon dioxide by using an indicator. An indicator is a substance that changes color in the presence of a certain chemical. Bromthymol blue is an indicator that changes color from blue to yellow in the presence of carbon dioxide.

Skills: observing, inferring, predicting, recording data

Objectives

In this laboratory, you will
- predict which of six unknown substances are living.
- use an indicator to test for the presence of carbon dioxide.

Note to teacher: The unknowns can include radish seeds, packaged yeast, brine shrimp eggs, sand, lentils, sawdust, sugar, salt, vermiculite, bean seeds.

Prelab Preparation

1. Review Section 2-1 Characteristics of Life and Section 2-2 Life Processes of Organisms.
2. Review how to use a compound microscope.
3. Review how to make a wet–mount slide.

Materials

5 unknown substances	compound microscope	scissors
bromthymol blue solution	microscope slides	plastic spoon
6 test tubes	cover slips	10 mL graduated cylinder
test tube rack	hand lens	marking pencil
6 Petri dishes		

Procedure

Part A: Observing the Unknowns

1. Use the hand lens to observe each unknown. **Caution: Do not taste any of the unknowns.**
2. Prepare a wet-mount of each unknown. Your teacher may cut thin samples of the unknown for you.
3. Use a compound microscopes to observe each of the unknowns.
4. Based on your observations, predict whether each unknown is living or nonliving. Record your predictions in Data Table 1.

Part B: Testing for Respiration

1. Use the marking pencil to label the six test tubes 1 through 6. Place the test tubes in the test tube rack.
2. Using test tubes 1–5, place 1g of each unknown into a test tube. Use test tube 6 as a control.
3. Add 3 mL of water to each test tube.
4. Add 3 mL of bromthymol blue to each test tube.
5. Observe the color of the solution in each test tube. Record your observations under "Day 1" in Data Table 1.
6. Put the test tubes and test tube rack in a place where they will not be disturbed until your next class period.
7. During your next class period, observe the color of the solution in each test tube. Record your observations under "Day 2" in Data Table 1.

Biology   Copyright ©   by Globe Book Company   9

Part C: Testing for Growth

1. Number six Petri dishes 1 through 6.
2. Using the bottom of a Petri dish as a guide, trace a circle on a paper towel. Cut out the circle. **Caution: Be careful when using scissors.**
3. Trace and cut out five more paper towel circles. Moisten each piece of paper towel circle with water.
4. Squeeze out any excess water and place one piece of paper towel into each Petri dish.
5. Use a plastic spoon to place a small amount of each unknown into Petri dishes 1-5. Use Petri dish 6 as a control.
6. Using a hand lens, observe each dish every day for four days. Be sure to keep the paper towels moist for the entire four days. Record your observations under "Growth" in Data Table 1.

Observations and Data

Data Table 1: Living or Nonliving

Unknown	Prediction (Living or nonliving)	Color Day 1	Color Day 2	Growth (Yes/No)
1	Answers will vary.			
2				
3				
4				
5				
control				

Analysis and Conclusions

1. a. Which unknown substances caused a color change in the bromthymol blue?
   Answers will vary.

   b. What does the color change indicate?
   Carbon dioxide was produced.

   c. What can you infer about the unknown substances that caused a color change?
   They are probably living.

2. a. Which of the unknowns grew during the four–day period?
   Answers will vary.

   b. What can you infer about the unknowns that grew?
   They are probably living.

3. a. Which of the unknown substances were living? Which were nonliving? Use your data to support your answer.
   Answers will vary.

   b. Were your predictions correct?
   Answers will vary.

Extension

Examine a drop of pond water under a compound microscope. Try to distinguish between living and nonliving things you see in the water sample. What guidelines could you use to distinguish between living and nonliving things? Record your observations in a data table.

Name _____ Class _____ Date _____

Chapter 3 — The Chemistry of Life

Laboratory 3-1 How are the properties of mixtures and compounds different?

Background Information

A mixture is a combination of two or more substances that are not chemically combined. The substances in a mixture can be separated physically. A compound is a combination of two or more substances that are chemically combined. The substances in a compound cannot be separated physically. The properties of a compound are different from the properties of the substances that reacted to form the compound.

Skills: observing, recording data, comparing and contrasting

Objectives

In this laboratory, you will
- choose the proper techniques to separate different mixtures.
- identify a combination of substances as a mixture or a compound.

Prelab Preparation

Review the following techniques for separating a mixture.
1. Filtration: Pour the mixture through a funnel lined with filter paper. The liquid will go through the funnel and the solid will remain on the filter paper. The setup for filtration is shown in Figure 1.

2. Evaporation: Gently heat a solution until all the liquid has evaporated. The solute will remain.

3. Separation by hand: Separate the particles of each substance by hand using a small paint brush. Use this technique if one of the substances has large particles.

4. Magnetism: Use a bar magnet to separate a magnetic substance from a nonmagnetic substance.

5. Dissolving: If one of the substances dissolves in water, add water to the mixture and then filter the mixture. The undissolved substance will remain on the filter paper. Evaporate the remaining liquid to recover the dissolved substance.

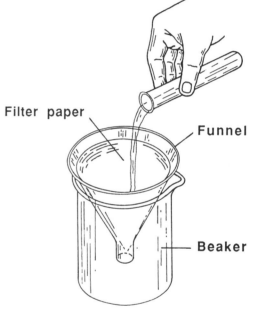

Figure 1: Filtration

Materials

filter paper	test tube holder	funnel
salt	small beaker	sand
graduated cylinder	sulfur	balance
iron filings	bar magnet	water
Bunsen burner	lead nitrate	3 test tubes
sodium iodide		

Biology   Copyright © by Globe Book Company   11

Procedure

1. Pour 10 mL of water into a test tube. Record the appearance of the water in Data Table 1.

2. Measure 3 g of salt and 3 g of sand. Record the appearance of the salt and the sand in Data Table 1.

3. Carefully add the salt and the sand to the water in the test tube.

4. Place your thumb over the mouth of the test tube and gently shake the tube for 30 seconds.

5. Using one or more of the techniques listed in the Prelab Preparation, separate the salt-sand-water mixture. Record the appearance of each substance after separation in Data Table 1.

6. Measure 3 g of sulfur. Record the appearance of the sulfur in Data Table 1.

7. Measure 3 g of iron filings. Record the appearance of the iron filings in Data Table 1.

8. Mix the sulfur and iron filings together on a piece of filter paper.

9. Using one or more of the techniques listed in the Prelab Preparation, separate the sulfur-iron mixture. Record the appearance of each substance after separation in Data Table 1.

10. Pour 10 mL of lead nitrate solution into a clean test tube. Record the appearance of the lead nitrate solution in Data Table 2.

11. Pour 10 mL of sodium iodide solution into another clean test tube. Record the appearance of the sodium iodide solution in Data Table 2.

12. Carefully pour the lead nitrate solution into the sodium iodide solution. Record the appearance of the combined solutions in Data Table 2.

Observations and Data

Data Table 1: Separating Mixtures

Substance	Appearance	
	Before Separation	After Separation
Water	**Students should record that**	
Salt	**each substance appears the**	
Sand	**same before and after**	
Sulfur	**separation.**	
Iron filings		

Name _____ Class _____ Date _____

Chapter 3   The Chemistry of Life   Laboratory 3-1

Data Table 2: Combining Two Substances

Substance	Appearance
Lead nitrate	Students should record that the appearance of the
Sodium iodide	combined solutions is different from the appearance
Lead nitrate and sodium iodide	of either of the original solutions.

Analysis and Conclusions

1. a. What techniques did you use to separate the salt-sand-water mixture?
   filtration and evaporation

   b. Why were you able to use these techniques?
   The sand did not dissolve in water and so could be separated by filtration. The salt dissolved in water and so could be separated by evaporation. (Note: The evaporated water could be recovered by condensation.)

2. a. What techniques did you use to separate the mixture of sulfur and iron filings?
   magnetism

   b. Why were you able to use these techniques?
   The iron filings are magnetic and can be separated from the nonmagnetic sulfur with a bar magnet.

3. Is the combination of lead nitrate and sodium iodide a mixture or a compound? How do you know?
   a compound, because the properties (or appearance) of the combined solutions are different from the properties of either of the original solutions

Biology

Extension

Combine equal amounts of iron filings, sand, salt, and water. Design an experiment to separate and recover each substance in the mixture. Describe your experiment in the space provided.

Name _____ Class _____ Date _____

Chapter 3   The Chemistry of Life

Laboratory 3-2 How can you identify inorganic and organic compounds?

Background Information

Most compounds that contain carbon are organic compounds. Compounds that do not contain carbon are inorganic compounds. You can identify organic and inorganic compounds by their properties. Most organic compounds will dissolve only in an organic solvent; they will not dissolve in water. Most inorganic compounds are made up of crystals, while organic compounds are not. Organic compounds tend to burn or char when heated, while inorganic compounds melt when heated.

Skills: observing, inferring, recording data

Objective

In this laboratory, you will
- determine if substances are organic or inorganic compounds.

Prelab Preparation

1. Review Section 3-3 Types of Compounds.

2. Review the procedure for lighting and using a Bunsen burner.

Materials

Bunsen burner
matches
hand lens
6 test tubes
test tube rack
filter paper
crucible
ring stand

tongs
plastic spoon
graduated cylinder
water
6 unknowns
paint thinner
clay triangle
ring

Note to teacher: The six unknowns are table sugar, salt, paraffin, calcium oxide, alum, and uncolored gelatin.

Caution: Paint thinner should be kept in closed containers at all times and should not be brought into the room until students are ready to use it.

Procedure

1. Place six pieces of filter paper on your work area. Number the papers 1 through 6.

2. Using a plastic spoon, place a small amount of the first unknown on the first piece of filter paper. Repeat with the remaining unknowns. Be sure to place each unknown on its proper numbered filter paper.

3. Observe unknown 1 with a hand lens. Determine if the particles of the unknown are the same size or different sizes. Record your observations in Data Table 1.

4. Repeat step 3 with the other five unknowns.

5. Again observe unknown 1 with a hand lens. Determine if the particles of the unknown have an irregular shape or a regular, crystal shape. Record your observations in Data Table 1.

6. Repeat step 5 with the other five unknowns.

7. **Caution: Be sure that no open flames are being used in the laboratory before proceeding with the following steps.** Set up six test tubes in a test tube rack. Add 10 mL of water to each test tube.

Biology   Copyright © by Globe Book Company   15

8. Place a small amount of unknown 1 in the first test tube, unknown 2 in the second test tube, and so on. Be sure to put the unknowns into the test tubes in the proper sequence.

9. Gently swirl the liquid in each test tube to help dissolve the unknowns.

10. Determine which of the unknowns are soluble in water. Record your observations in Data Table 1.

11. Clean and rinse each test tube thoroughly.

12. **Caution: Extinguish all open flames. Wear your apron and goggles.** Add 10 mL of paint thinner to each test tube.

13. Repeat steps 8 and 9.

14. Determine which of the unknowns are soluble in paint thinner. Record your observations in Data Table 1.

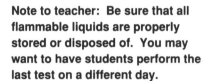

Note to teacher: Be sure that all flammable liquids are properly stored or disposed of. You may want to have students perform the last test on a different day.

15. Clean and rinse each test tube thoroughly. Follow your teacher's instructions for proper disposal of the paint thinner.

16. Set up a ring stand, ring, clay triangle, and Bunsen burner as shown in Figure 1.

17. Place a small amount of your unknown into a crucible.   **Note to teacher: Divide the class into six groups; assign one unknown to each group.**

18. Place the crucible with your unknown on the ring over the Bunsen burner.

19. Light the Bunsen burner. **Caution: Follow the proper procedure for lighting a Bunsen burner.**

20. Heat the unknown and observe if it melts, chars, or burns. Record your observations in Data Table 1.

21. Combine your observations with the other groups in your class to complete Data Table 1.

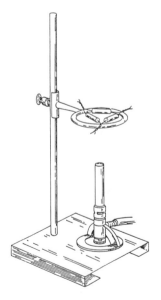

Figure 1: Experimental Set-Up

Name _____ Class _____ Date _____

Chapter 3   The Chemistry of Life   Laboratory 3-2

Observations and Data

Data Table 1: Inorganic and Organic Compounds

Unknown	Particles		Solubility		Reaction to Heat
	Size	Shape	Water	Paint Thinner	
1	Answers will				
2	vary.				
3					
4					
5					
6					

1. Which substances showed a crystal structure when observed with a hand lens?
   Answers will vary.

2. Which substances dissolved in water?
   Answers will vary.

3. Which substances dissolved in paint thinner?
   Answers will vary.

4. Which substances melted when heated?
   Answers will vary.

5. Which substances charred or burned when heated?
   Answers will vary.

Analysis and Conclusions

Which unknown substances do you think are organic compounds? Which are inorganic? Use your data to support your answers.

Answers will vary. Paraffin, sugar, and gelatin are organic; salt, calcium oxide, and alum are inorganic.

Biology

Extension

Many substances have the ability to conduct electricity when dissolved in water. Design an experiment to answer the following question: Which substances are better conductors of electricity when dissolved in water — organic compounds or inorganic compounds? Describe your experiment in the space provided.

Name _____ Class _____ Date _____

Chapter 3 The Chemistry of Life

Laboratory 3-3 How can you identify the compounds found in living things?

Background Information

All living things are made up of cells. Cells contain organic and inorganic compounds that are needed to carry on an organism's life processes. Some of the compounds needed by living things are water, salts, sugars, starches, fats, and proteins.

Skills: observing, inferring, recording data, analyzing

Note to teacher:
See p. T-xvi for preparation of unknown solution.

Objectives

In this laboratory, you will
- use indicators and chemical tests to analyze an unknown solution.
- determine the chemical compounds found in living things.

Prelab Preparation

Review the following tests for chemical compounds.

1. Test for water: Heat a small sample of the unknown in a test tube over a Bunsen burner. Moisture on the inside of the tube near the top indicates that water is present.

2. Test for sugar: Add Benedict's solution to a small amount of the unknown in a test tube. Heat gently over a Bunsen burner. A color change to green, yellow, orange, or brick red indicates that sugar is present.

3. Test for starch: Add Lugol's solution to a small amount of the unknown in a test tube. A color change to blue-black indicates the presence of starch.

4. Test for fats: Rub a small amount of the unknown onto a piece of plain brown paper. A greasy spot that does not dry up indicates the presence of fats.

5. Test for proteins: Add Biuret solution to a small amount of the unknown in a test tube. A color change from light blue to purple indicates the presence of protein.

6. Test for salt: Filter the unknown into a shallow dish and allow the filtered liquid to evaporate. Examine the solid remaining in the dish with a hand lens for the presence of salt crystals.

Materials

unknown solution	Bunsen burner	6 test tubes
test tube rack	test tube holder	funnel
filter paper	shallow dish	brown wrapping paper
hand lens	dropper	Benedict's solution
Lugol's solution	Biuret solution	graduated cylinder

Procedure

1. Place six clean, dry test tubes in a test tube rack. Number the tubes from 1 to 6.

2. Add 10 mL of the unknown solution to each of the test tubes.

3. Following the procedures described in the Prelab Preparation, test each sample of the unknown solution for the presence of one of the compounds listed. Test the solution in test tube 1 for water, in test tube 2 for sugar, and so on. Be sure to perform the tests in the sequence listed. Record your results in Data Table 1.

Biology Copyright © by Globe Book Company 19

Observations and Data
Data Table 1: Test Results

Test Tube	Water	Sugar	Starch	Fat	Protein	Salt
1	√					
2		√				
3			√			
4				√		
5					√	
6						√

Analysis and Conclusions

1. An indicator is a substance that changes color in the presence of a certain chemical. What indicators did you use in this laboratory?
   Benedict's solution, Lugol's solution, Biuret solution

2. Which chemical compound was each indicator used to identify?
   Benedict's solution: sugar

   Lugol's solution: starch

   Biuret solution: protein

3. How can you identify the compounds found in living things?
   Answers will vary. Students should indicate that careful observation of the results of chemical tests using indicators and physical tests can be used to identify chemical compounds.

Extension

Repeat the tests used in this laboratory with samples of foods such as egg white, tuna fish, bacon, milk, peanuts, cottage cheese, mayonnaise, potato, carrot, peanut butter, breakfast cereal, fruit, and uncooked ground beef. Record your results in a data table.

Name _____ Class _____ Date _____

Chapter 4   Cells

Laboratory 4-1   How do plant cells and animal cells differ in structure and function?

Background Information

All living things are made up of cells. The structure of a plant or animal cell is related to its function. Plant cells and animal cells have different shapes and contain different organelles. You can identify plant cells and animal cells by observing their shape and structure with a microscope.

Skills: observing, inferring, classifying, organizing data

Objectives

In this laboratory, you will
- prepare and observe slides of three different plant cells.
- test for the presence of starch in three different plant parts.
- observe prepared slides of animal cells.
- compare the structure of plant cells and animal cells.

Prelab Preparation

1. Review Section 4-2 Cell Structure and Function.

2. Review the procedure for using a compound microscope.

3. Define the term "chromoplast."

   an organelle that makes and stores red, orange, or yellow pigment

Materials

compound microscope	iodine stain	microscope slides
cover slips	paper towels	*Elodea* leaf
forceps	tomato slice	dropper
potato slice	razor blade	prepared slide of human skin cells

Procedure

Part A: Structure of Plant Cells

1. Place a thin *Elodea* leaf on a clean, dry slide.

2. Place a drop of water on the leaf.

3. Slowly lower a cover slip onto the leaf. Try to avoid air bubbles.

4. Observe your wet-mount slide with the high power of a microscope.

5. Draw what you see in Plate 1. Label the cell parts you observed.

6. Record your observations in Data Table 1.

7. Peel a small section of skin from a tomato slice.

Biology   Copyright ©   by Globe Book Company

8. Using a razor blade, carefully scrape the pulp from the inner surface of the skin. **Caution: Be careful when using a razor blade.**

9. Place the inner surface of the tomato skin on a slide and make a wet mount.

10. Observe the slide under high–power magnification.

11. Draw what you see in Plate 2. Label the cell parts you observed.

12. Record your observations in Data Table 1.

13. Using a razor blade, carefully slice a thin section of potato. **Caution: Be careful when using a razor blade.**

14. Place a slice of potato on a slide and make a wet mount.

15. Observe the slide under high–power magnification.

16. Draw what you see in Plate 3. Label the cell parts you observed.

17. Record your observations in Data Table 1.

Part B: Testing for Starch

1. To test for starch, place an *Elodea* leaf, a small piece of tomato, and a small piece of potato on a paper towel.

2. Place a drop of iodine on the leaf, the piece of tomato, and the piece of potato. A blue–black color indicates that starch is present.

3. Record your observations in Data Table 1.

Part C: Structure of Animal Cells

1. Observe a prepared slide of human skin cells under high–power magnification.

2. Draw what you see in Plate 4. Label the cell parts you observed.

3. Record your observations in Data Table 1.

Observations and Data

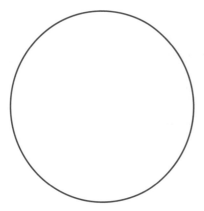

Plate 1: *Elodea* Leaf

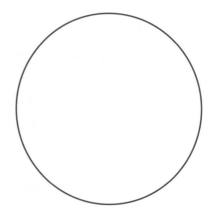

Place 2: Tomato Skin

Name_____ Class_____ Date_____

Chapter 4 Cells

Laboratory 4-1

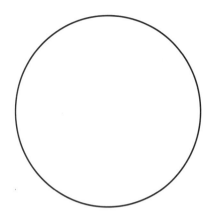
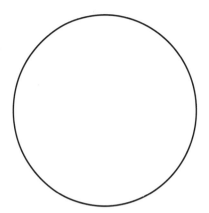

Plate 3: Potato Slice   Plate 4: Human Skin

Data Table 1: Cell Parts

Cells	Cell Parts Observed	Starch Present	
		Yes	No
Elodea	cytoplasm, nucleus, cell membrane, cell wall, vacuoles, chloroplasts		X
Tomato	cytoplasm, nucleus, cell membrane, cell wall, vacuoles, chromoplasts		X
Potato	cytoplasm, nucleus, cell membrane, cell wall, vacuoles	X	
Human skin	cytoplasm, nucleus, cell membrane, vacuoles	—	—

Note to teacher: The cell membrane of the plant cells may not be visible with the light microscope.

Analysis and Conclusions

1. How are the *Elodea*, tomato, and potato cells alike?
   All three have cytoplasm, nuclei, cell membranes, cell walls, and vacuoles.

2. How are the three plant cells different?
   Answers will vary. *Elodea* cells have chloroplasts; tomato cells have chromoplasts. Their shapes are different.

Biology   Copyright ©   by Globe Book Company   23

3. Which kind of plant cell contained chloroplasts? Explain your answer.

   Elodea leaf cells contained chloroplasts; leaves are involved in photosynthesis.

4. Which kind of plant tested positive for starch? Explain your answer.

   The potato tested positive for starch; potatoes store food in the form of starch.

5. a. How are animal cells similar to plant cells?

   Answers will vary. Both have cytoplasm, nuclei, cell membranes, and vacuoles.

   b. How are animal cells different from plant cells?

   Answers will vary. Animal cells have no cell walls and no chloroplasts.

   They have smaller vacuoles; they are more irregular in shape and contain

   centrioles.

6. a. Why do you think all cells do not have the same shape?
   b. Why do they not have the same organelles?

   Different cells have different functions. The shape and structure of a cell are adapted to its function.

Extension

Examine unlabeled, prepared slides of different animal and plant cells. Based on your observations, identify each unknown slide as showing plant cells or animal cells.

Name _____ _____ Class _____ Date _____

Chapter 4 Cells

Laboratory 4-2 How do mitosis and meiosis compare?

Background Information

Like all organisms, a human body develops from a single cell. As you grow, your body is continually producing new cells. The process by which new body cells are produced is called mitosis. Through this process your body makes new cells to replace worn-out cells. Most organisms also undergo a second kind of cell division called meiosis. Meiosis is the process by which sex cells, or eggs and sperm, are formed.

Skills: observing, classifying, modeling

Objectives

In this laboratory, you will
- observe prepared slides of onion root tip cells and identify the stages of mitosis.
- use a model to identify the stages of meiosis.

Prelab Preparation

1. Review Section 4–4 Cellular Reproduction.

2. Review the procedure for using the compound microscope.

3. Review Figure 4–14, Mitosis, and Figure 4–15, Meiosis.

Materials

compound microscope	prepared slide of onion root tip	scissors
glue	tape	construction paper
yarn	pipe cleaners	

Procedure

Part A: Observing Mitosis

1. Place a prepared slide of an onion root tip on the stage of a compound microscope.

2. Examine the slide with the low power of the microscope. Then switch to high power.

3. Look at Figure 4–14 on page 69 of your textbook.

4. Carefully examine your slide and compare what you see with the stages of mitosis shown in Figure 4–14. Find cells that look like the stages of mitosis shown in Figure 4–14. Draw what you see in Plate 1.

5. Label the following structures on your drawings: chromosomes, nuclear membrane, cytoplasm, spindle fibers, daughter cells.

Part B: Making a Model of Meiosis

1. Look at Figure 4–15 on page 70 of your textbook.

2. Use construction paper, yarn, and pipe cleaners to make a model showing the stages of meiosis. Use Figure 4–15 as a guide.

3. Label the following structures on your model; nucleus, chromosomes, spindle fibers, nuclear membrane.

Biology   Copyright ©   by Globe Book Company   25

Observations and Data

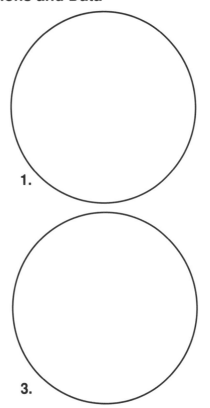
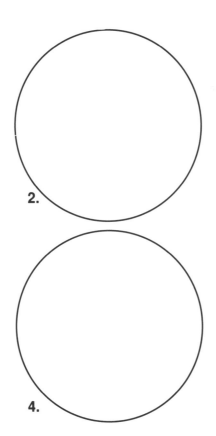

Plate 1: Stages of Mitosis

1. How many daughter cells are formed as a result of mitosis?

   two

2. How many sex cells are formed as a result of meiosis?

   four

Analysis and Conclusions

1. How is mitosis similar to meiosis?

   In both processes, chromosomes are duplicated, cells divide, and new cells are formed.

2. How is mitosis different from meiosis?

   As a result of mitosis, identical daughter cells with complete sets of chromosomes are formed. As a result of meiosis, special sex cells with half the number of chromosomes as in body cells are formed.

Extension

Use reference materials to find out the names of the different stages in mitosis and meiosis. Add the correct names for each stage to your drawings and model.

Name _____ Class _____ Date _____

Chapter 5 — Classification

Laboratory 5-1 How can a dichotomous key be used to identify an organism?

Background Information

People organize things to make it easier to identify them. For example, compact discs, audio casettes, and records are usually grouped in three different sections in an audio store. Scientists also group, or classify, organisms based on similarities and differences. One way to group organisms is by using a dichotomous key. The term "dichotomous" means "divided into two parts." A dichotomous key is designed to separate a group of organisms into two smaller groups. These groups are then separated into two smaller groups, and so on. Each step in the key involves a yes or no question about the organism's appearance.

Skills: observing, classifying, inferring

Objectives

In this laboratory, you will
- identify animals based on their appearance.
- use a dichotomous key to identify leaves.

Prelab Preparation

Review Section 5-1 Classifying Living Things and Section 5-2 Modern Taxonomy.

Materials

paper   pencil

Procedure

1. Look at the animals shown in Figure 1.

2. Complete Data Table 1 based on your observations of the animals in Figure 1.

Figure 1. Six Different Animals

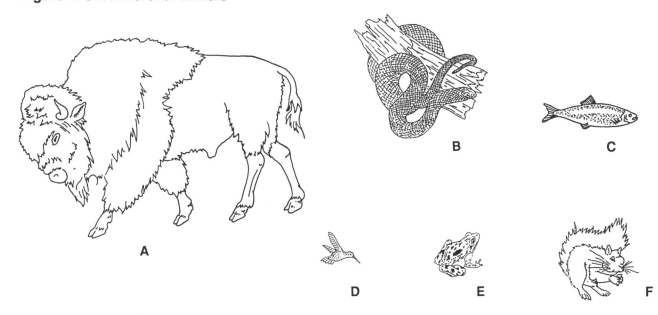

Biology — Copyright © by Globe Book Company — 27

3. Look at the leaves shown in Figure 2.

Figure 2. Six Different Leaves

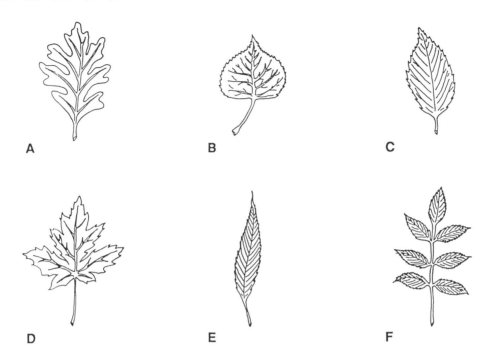

4. Use the dichotomous key in Figure 3 to identify each of the leaves in Figure 2. Record your results in Data Table 2.

Figure 3. Dichotomous Key for Leaves

	Questions	Yes	No
1.	Is the leaf blade made up of small leaflets?	The leaf is a green ash.	Go to 3.
2.	Is the leaf blade cut into lobes?	Go to 4.	Go to 5.
3.	Is the leaf blade long and narrow?	The leaf is a black willow.	Go to 2.
4.	Are the lobes pointed?	The leaf is a silver maple.	The leaf is a white oak.
5.	Is the leaf heart shaped?	The leaf is a cottonwood.	The leaf is an American elm.

Name_____ Class_____ Date_____

Chapter 5   Classification

Laboratory 5-1

Observations and Data

Data Table 1: Animals

Animal	Fins	Wings	Legs 2	Legs 4	Hooves	Claws	Scales	Feathers	Fur	Smooth Skin
A				√	√				√	
B							√			
C	√						√			
D		√	√					√		
E				√						√
F				√		√			√	

Data Table 2: Leaves

Leaf	Identification
A	white oak
B	cottonwood
C	American elm
D	silver maple
E	black willow
F	green ash

Analysis and Conclusions

1. Based on your observations in Data Table 1, match each animal listed with the correct letter in Figure 1.

   Fish   __C__

   Frog   __E__

   Snake   __B__

   Hummingbird   __D__

   Squirrel   __F__

   Bison   __A__

Biology   Copyright © by Globe Book Company   29

2. Classify each of the animals in Figure 1 in one of the following groups:

Fish	__C__	Bird	__D__
Amphibian	__E__	Mammal	__A, F__
Reptile	__B__		

3. How did the dichotomous key help you to identify the leaves in Figure 2?

By answering a series of questions about the appearance of the leaves, it is possible to identify each leaf.

Extension

Find photographs of at least 10 different flowers. Develop a dichotomous key based on the number and arrangement of the petals to identify each flower.

Name _____ Class _____ Date _____

Chapter 6   Viruses and Monerans

Laboratory 6-1 Where are bacteria found in your environment?

Background Information

Bacteria are classified in the Kingdom Monera. Most bacteria grow best in a dark, moist environment where the air temperature is between 25 and 40 degrees Celsius. However, bacteria can remain inactive, or dormant, for a long time if the conditions for their growth and survival are not present. When conditions improve, the bacteria again become active.

Skills: observing, comparing, inferring

Objective

Have students prepare the Petri dishes with nutrient agar before the laboratory. Store in the refrigerator upside down.

In this laboratory, you will
- observe the growth of bacteria taken from different locations.

Prelab Preparation

Review Section 6-2 Monerans.

Materials

2 Petri dishes   hand lens   nutrient agar
marking pencil   cotton swabs

Procedure

1. Turn a Petri dish upside down. With a marking pencil, divide the bottom of the dish into four quarters. Number each quarter as shown in Figure 1.

2. Turn the Petri dish right-side up. Label the lid of the Petri dish "A."

3. Repeat step 1 with the second Petri dish. Label the lid of this dish "B."

4. Rub a cotton swab over a surface, such as your desk.

Figure 1

5. Carefully lift the lid of Petri dish A. Gently rub the cotton swab back and forth over the surface of the agar in section 1. See Figure 2. Do not touch the agar with your fingers. Quickly lower the lid. Dispose of the swab.

6. Repeat steps 5 and 6 for three other surfaces. Use a clean swab for each location.

7. Put both Petri dishes A and B in a warm, dark place where they will not be disturbed.

8. Observe the Petri dishes every day for 3 to 5 days. Do not open the Petri dishes. Record your observations.

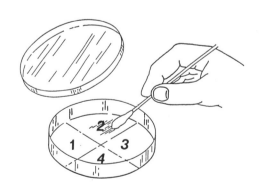

Figure 2

9. After 3 to 5 days, draw what you see in Petri dish A in Plate 1. Draw what you see in Petri dish B in Plate 2.

Biology   Copyright © by Globe Book Company

Observations and Data

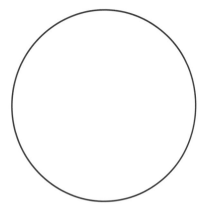

Plate 1: Petri dish A

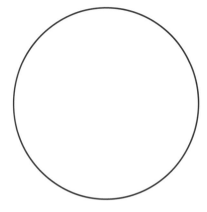

Plate 2: Petri dish B

Analysis and Conclusions

1. What was the purpose of Petri dish B?

   to act as a control

2. From which location did you observe the greatest bacterial growth?

   Answers will vary.

3. Compare your results with other members of your class. Which location tested produced the greatest amount of bacterial growth?

   Answers will vary.

4. How many different kinds of bacteria were you able to grow? Use your observations to support your answer.

   Answers will vary.

5. Why did you observe bacterial growth on the nutrient agar, but not on the surfaces you tested?

   Bacteria grew on the nutrient agar in the Petri dish because all of the necessary conditions for growth were present.

Extension

In addition to nutrient agar, what substances can be used as a medium for the growth of bacteria? Design an experiment to test several different substances, such as beef or chicken broth, for bacterial growth. How could you observe the growth of bacteria in the broth without using a microscope?

Name _____ Class _____ Date _____

Chapter 7 — Protists

Laboratory 7-1 How do different kinds of protozoans compare?

Background Information

Protozoans are often called the animal-like protists. Most protozoans cannot make their own food but must move about to find food. Biologists classify protozoans based on how they move.

Skills: observing, comparing, analyzing

Objective

In this laboratory, you will
- observe two different protozoans and compare their structure and method of movement.

Prelab Preparation

1. Review the procedure for using a compound microscope.

2. Review how to make a wet-mount slide.

3. Review Section 7-2 Protozoans.

Materials

prepared slides of amoebas and paramecia
live cultures of amoebas and paramecia
compound microscope
cotton

cover slips
microscope slides
dropper

Procedure

1. Place a prepared slide of amoebas on the stage of a compound microscope.

2. Observe an amoeba using the high power of the microscope. Locate the cell membrane, nucleus, cytoplasm, and a pseudopod.

3. Sketch what you see in Plate 1. Label cell membrane, nucleus, cytoplasm, pseudopod.

4. Repeat steps 1 and 2 with a prepared slide of paramecia. Sketch what you see in Plate 2. Label pellicle, nucleus, cytoplasm, and cilia.

5. To observe movement in protozoans, use live cultures of amoebas and paramecia.

6. Place a few strands of cotton on a clean, dry slide.

7. Use a dropper to obtain a sample of amoeba culture from the bottom of the container. Make a wet-mount slide of the amoeba sample.

8. Locate a single amoeba and observe how it moves. Record your observations.

9. Repeat steps 6 to 8 with the paramecium culture.

Biology   Copyright © by Globe Book Company

Observations and Data

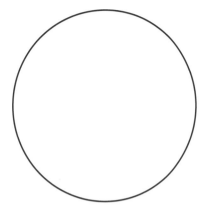

Plate 1: Amoeba

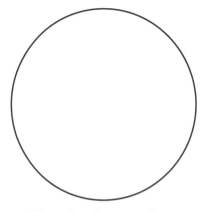

Plate 2: Paramecium

1. Describe how an amoeba moves.
   by means of pseudopods, or fingerlike projections of cytoplasm

2. Describe how a paramecium moves.
   in a spiralling motion, propelled by beating cilia

Analysis and Conclusions

1. What was the purpose of the cotton strands on your slides of the amoeba and paramecium cultures?
   to slow the organisms down

2. How are the structures of an amoeba and a paramecium related to the way they move? Use your data to support your answer.
   Answers will vary. Amoebas have an irregular shape and move by amoeboid motion, using pseudopods.

   Paramecia have a definite shape and move by means of beating cilia.

Extension

Obtain a sample of pond water. Locate at least two protozoans other than amoebas and paramecia. Sketch each protozoan and describe the way it moves. Use your textbook or other reference materials to identify each protozoan.

Name _____ Class _____ Date _____

Chapter 7 | Protists

Laboratory 7-2   How do different kinds of algae compare?

Background Information

Spirogyra, *Volvox*, and *Euglena* are three kinds of algae. *Spirogyra* form filaments or chains of individual cells. The filaments have a slimy, jellylike covering. *Volvox* form ball-shaped colonies of single cells. The cells of the colony are held together by a jellylike material. The colony moves by means of flagella. *Euglena* are single-celled algae that move by means of flagella. Each *Euglena* has a light-sensitive organelle called an eyespot.

Skills: observing, comparing, inferring

Objective

In this laboratory, you will
- observe three kinds of algae and compare their structure.

Prelab Preparation

1. Review the procedures for using a compound microscope.

2. Review Section 7-3 Algae.

Materials

prepared slides of *Spirogyra*, *Volvox*, and *Euglena*
compound microscope

Procedure

1. Place a prepared slide of *Spirogyra* on the stage of a compound microscope.

2. Observe the slide using the low power of the microscope. Locate the cell walls, chloroplasts, and nuclei.

3. Sketch what you see in Plate 1. Label chloroplast, cell walls, and nucleus.

4. Observe a prepared slide of *Volvox*. Sketch what you see in Plate 2.

5. Observe a prepared slide of *Euglena*. Sketch what you see in Plate 3. Label chloroplast, flagellum, nucleus, and eyespot.

Biology Copyright © by Globe Book Company 35

Observations and Data

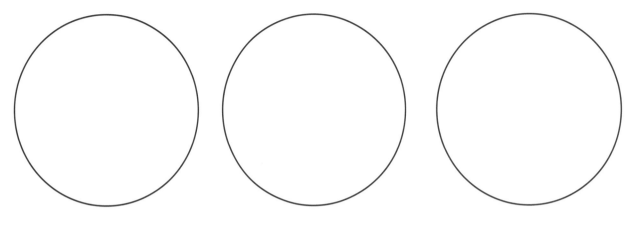

Plate 1: *Spirogyra*　　Plate 2: *Volvox*　　Plate 3: *Euglena*

1. Which of the three algae have chloroplasts?

    All three have chloroplasts.

2. Which of the three algae have flagella?

    Volvox and Euglena　(Note to teacher: Flagella may not be visible on prepared slides of *Volvox*.)

3. What structure is present only in *Euglena*?

    eyespot

Analysis and Conclusions

1. At one time, *Euglena* were classified as protozoans. What animallike characteristics of *Euglena* might cause biologists to classify them as protozoans?

    flagellum and eyespot

2. Do you think *Spirogyra* are able to move in the same way as *Volvox* or *Euglena*? Use your data to support your answer.

    No, because Volvox and Euglena have flagella, while Spirogyra do not.

Extension

Using a culture of live *Euglena*, design an experiment to test their reaction to light. In what direction do the *Euglena* swim? Is the same end always facing toward or away from the light? Record your observations in a data table.

Name _____ Class _____ Date _____

Chapter 8 — Fungi

Laboratory 8-1 Can the growth of molds be slowed or stopped?

Background Information
Molds are living things that are part of the Kingdom Fungi. Molds reproduce by means of spores. If a mold spore lands on food, it will begin to grow. The mold sends its rootlike rhizoids into the food and begins to digest the food. Unless the growth of molds is slowed or stopped, molds can destroy many kinds of foods eaten by people.

Skills: observing, comparing and contrasting, analyzing

Objectives
In this laboratory, you will
- observe the growth of molds on different kinds of foods.
- test different substances to determine which can slow or stop the growth of molds.

Prelab Preparation
1. Review Section 8-2 Molds.

2. What is the definition of the term "dehydrate"?

   <u>to remove water from a substance</u>

Materials

15 baby food jars with lids	bread	crackers
oatmeal	orange	rice
2 household cleaners	bleach	ammonia
water	hand lens	plastic knife
plastic spoon	marking pencil	dropper
cotton swabs		

Procedure

Part A: Growing Molds

1. Label 10 baby food jars 1 to 10.

2. Put small pieces of bread into jars 1 and 2.

3. Put several pieces of cracker into jars 3 and 4.

4. Use the knife to cut an orange into quarters. **Caution: Be careful when using the knife.** Put a piece of orange into jars 5 and 6.

5. Put a spoonful of rice into jars 7 and 8.

6. Put a spoonful of oatmeal into jars 9 and 10.

7. Add 20 drops of water to jars 2, 4, 6, 8, and 10.

8. Let the jars stand open for 5 minutes. After 5 minutes, put the lids on the jars.

9. Put the jars in a warm, dark place where they will not be disturbed for three days.

10. Observe the jars every day for three days. Look for any growth of molds. Notice any difference in the color of the molds. Use a hand lens, but do not remove the lids from the jars.

Biology   Copyright © by Globe Book Company   37

11. After three days, record your observations in Data Table 1.

Part B: Slowing or Stopping the Growth of Molds

1. Label five baby food jars 1 to 5.

2. Put a piece of bread into each jar.

3. Add 20 drops of water to each jar.

4. Using a cotton swab, remove a sample of mold that you grew in Part A. Rub the swab over the bread pieces in jars 1 to 5.

6. Add 10 drops of bleach to jar 1. **Caution: Be careful when adding bleach. Do not inhale fumes from the bleach or spill any on your skin.**

7. Add 10 drops of ammonia to jar 2. **Caution: Be careful when adding ammonia. Do not inhale ammonia fumes or spill any on your skin. Do not mix bleach and ammonia.**

8. Add 10 drops of household cleaner to jar 3.

9. Add 10 drops of a different household cleaner to jar 4.

10. Add nothing to jar 5.

11. Put each jar in a warm, dark place where they will not be disturbed for five days.

12. After three days, observe each jar for evidence of mold growth. Record your observations in Data Table 2.

13. Continue to observe each jar every day for the next three days. Record your observations in Data Table 2.

Observations and Data

Data Table 1: Growing Molds

Jar	Mold Growth (yes or no)
1 (bread)	no
2 (bread + water)	yes
3 (cracker)	no
4 (cracker + water)	yes
5 (orange)	no
6 (orange + water)	yes
7 (rice)	no
8 (rice + water)	yes
9 (oatmeal)	no
10 (oatmeal + water)	yes

Name _____ Class _____ Date _____

Chapter 8 Fungi
Laboratory 8-1

Data Table 2: Slowing or Stopping Mold Growth

Jar	Day 3	Day 4	Day 5	Day 6
1 (bleach)	no growth	no growth	no growth	no growth
2 (ammonia)	no growth	no growth	no growth	some growth
3 (cleaner)	no growth	no growth	some growth	some growth
4 (cleaner)	no growth	no growth	some growth	some growth
5 (nothing)	no growth	some growth	some growth	some growth

1. a. In which of the jars listed in Data Table 1 did you observe mold growth?
   2, 4, 6, 8, 10

   b. In which jars did you observe no growth?
   1, 3, 5, 7, 9

2. How many different kinds of molds did you observe?
   Answers will vary.

Analysis and Conclusions

1. What conditions are necessary for the growth of molds? Use your data to support your answer.
   moist, warm

2. In storing food, what conditions would prevent the growth of molds?
   cool, dry

3. How could you identify different kinds of molds in Part A?
   by their different colors

4. a Which of the substances tested in Part B slowed the growth of molds?
   Answers will vary. Students should observe that bleach is most effective in slowing mold growth, followed by household cleaners and ammonia.

Biology   Copyright ©   by Globe Book Company   39

b. What was the purpose of jar 5 in Part B?

to serve as a control

5. Most foods contain preservatives to prevent the growth of molds, but many dehydrated foods do not contain preservatives. What prevents the growth of molds in dehydrated foods?

lack of water

Extension

Which preservative is most effective in slowing or stopping the growth of molds on food? Design an experiment to answer this question. Record your results in a data table. Suppose the use of chemical preservatives in food was banned. What effect would this action have on consumers?

Name _____ Class _____ Date _____

Chapter 8   Fungi

Laboratory 8-2   How do yeast cells reproduce and carry on respiration?

Background Information

Yeasts are microscopic, single–celled fungi. They usually reproduce asexually by budding. Unlike most organisms, yeasts usually do not obtain energy through respiration. Instead, they obtain energy through a form of respiration called fermentation. In the process of fermentation, sugar is broken down into alcohol and carbon dioxide.

Skills: hypothesizing, inferring, observing

Objectives

In this laboratory, you will
- observe budding in yeast cells.
- use an indicator to identify carbon dioxide produced by yeast cells.

Sugar Solution: Add 1g sucrose to 50 mL tap water; stir. Yeast suspension: Add a pinch of dry Baker's yeast to 50 mL tap water.

Prelab Preparation

1. Review Section 8–3 Sac Fungi.
2. Review the procedure for using a compound microscope.
3. Review the function of an indicator.

Materials

compound microscope	microscope slide	cover slip
bromthymol blue solution	5 test tubes	test tube rack
graduated cylinder	dropper	marking pencil
yeast suspension	sugar solution	

Procedure

Part A: Budding

1. Place a drop of yeast suspension on a clean, dry microscope slide. Slowly lower a cover slip over the drop. Be sure not to trap any air bubbles under the cover slip.

2. Examine the slide under the low power of the microscope. Then switch to high power.

3. Refer to Figure 1. Search your slide to locate budding yeast cells as shown in Figure 1.

4. Draw what you see in Plate 1. Label a bud.

Figure 1: Budding Yeast Cells

Part B: Fermentation

1. Label three test tubes A, B, and C. Place the test tubes in a test tube rack.

2. Add 3 mL of bromthymol blue solution to each test tube.

3. Add a dropper full of yeast suspension to test tubes B and C.

4. Add a dropper full of sugar solution to test tube C.

5. Observe the color of the solution in each test tube. Record your observations under "Day 1" in Data Table 1.

6. Put the test tubes and test tube rack in a warm, dark place where they will not be disturbed until your next class period.

Biology   Copyright ©   by Globe Book Company   41

7. During your next class period, observe the color of the solution in each test tube. Record your observations under "Day 2" in Data Table 1.

Observations and Data

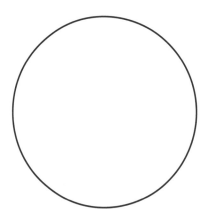

Plate 1: Budding Yeast Cells

Data Table 1: Fermentation in Yeast

Test Tube	Color	
	Day 1	Day 2
A (bromthymol blue)	Answers will vary.	
B (bromthymol blue + yeast)		
C (bromthymol blue + yeast + sugar)		

1. In which test tubes did the bromthymol blue not change color after 24 hours?
   test tube A

2. In which test tubes did the bromthymol blue change color?
   test tubes B and C

Analysis and Conclusions

1. What was the purpose of test tube A in this laboratory?
   to serve as a control

2. What caused the bromthymol blue to change color?
   carbon dioxide produced by the yeast cells

3. What was the purpose of the sugar solution in this laboratory?
   to serve as a food source that was broken down into alcohol and carbon dioxide during the fermentation process

Extension

Design an experiment to find out if yeast cells can break down saccharine in the same way they break down sugar during the process of fermentation. Record your results in a data table.

Name _____ Class _____ Date _____

Chapter 9 — Spore Plants

Laboratory 9-1 How does the amount of available moisture and light affect the growth of mosses?

Background Information

Mosses are nonvascular plants that rarely grow more than a few centimeters tall. Because they are so small, mosses do not need much soil. They can often be found growing on rocks or in the cracks of sidewalks. Their rootlike parts, or rhizoids, anchor mosses in the soil and absorb water.

Skills: observing, measuring, analyzing

Objective

In this laboratory, you will
- observe how moisture and light affect the growth of moss plants.

Prelab Preparation

1. Review Section 9-2 Mosses, Liverworts, and Hornworts.

2. Review how to find an average.

Materials

moss plants	potting soil	water
4 paper cups	scissors	hand lens
metric ruler		

Procedure

1. Carefully cut each of the four paper cups in half. **Caution: Be careful when using scissors.** Number the cups 1 to 4.

2. Place 3 cm of potting soil into each cup. Do not pack the soil too firmly. Moisten the soil.

3. Place a piece of moss plant into each cup. Observe the moss with a hand lens. Use a metric ruler to measure the average height of the moss in each cup. Record your measurements in Data Table 1.

4. Put cup 1 on a windowsill where it will receive direct sunlight for several hours each day.

5. Put cup 2 in a shady spot where it will not receive any direct or indirect sunlight.

6. Keep the moss plants moist by watering them once every day. Be sure to add the same amount of water to each cup.

7. Put cups 3 and 4 in a shady spot where they will not receive any sunlight. Water the moss in cup 3 once every day. Water the moss in cup 4 once every other day.

8. Observe and measure the average height of the moss plants in each of the four cups every day for five days. Record your observations in Data Table 1.

Biology — Copyright © by Globe Book Company

Observations and Data

Data Table 1: Moss Plants

Cup	Average Height (cm)				
	Day 1	Day 2	Day 3	Day 4	Day 5
1 (light)					
2 (shade)			Answers will		
3 (moist)			vary.		
4 (dry)					

1. Which cups showed the most growth after five days?
   cups 2 and 3

2. Which cups showed the least growth?
   cups 1 and 4

Analysis and Conclusions

1. What were the two variables tested in this laboratory?
   moisture and light

2. a. What conditions are most favorable for the growth of moss plants? Use your data to support your answer.
   moisture and shade

   b. Why do you think mosses require these conditions for growth?
   Mosses require moisture because they have no vascular tissue to transport water; water must diffuse directly from cell to cell. Mosses grow better in shade than in sunlight because they lose less water by evaporation in shade than in sunlight.

Extension

Design an experiment to find out how much water a moss plant can hold. Measure the mass of a well-watered moss plant. Then allow the moss to dry out completely. Measure the mass of the dry moss plant. Calculate the amount of water held by the moss. Record your results in a data table.

Name_____ Class _____ Date _____

Chapter 10 | Seed Plants

Laboratory 10-1 How can a dichotomous key be used to identify different kinds of conifers?

Background Information

Conifers are cone-bearing seed plants. They can be identified by their needlelike leaves and the cones scattered among their branches. Four kinds of conifers are pines, spruce, firs, and cedars. Each of these conifers can be identified by the shape of its needles.

Skills: observing, classifying, inferring

Objective

In this laboratory, you will
- examine twigs from four different conifers and use a dichotomous key to identify each conifer.

Prelab Preparation

Review Section 10-2 Gymnosperms.

Materials

twigs of pine, spruce, fir, and cedar
metric ruler

paper
pencil

Procedure

1. Select one of the conifer twigs. This will be specimen A.

2. Use the dichotomous key in Figure 1 to identify specimen A. Record your results in Data Table 1.

3. Repeat steps 1 and 2 for the remaining three conifer twigs. Refer to these twigs as specimens B, C, and D in Data Table 1.

Observations and Data

Data Table 1: Conifers

Specimen	Identification
A	Answers will vary.
B	
C	
D	

Biology   Copyright © by Globe Book Company   45

Figure 3. Dichotomous Key for Leaves

Questions	Yes	No
1. Does the twig have leaves that look like needles?	Go to 2.	Go to 8.
2. Are the needles in groups of two or more?	Go to 3.	Go to 4.
3. Are the needles at least 5 cm long, and two to five in a bundle?	This is a pine.	Go to 5.
4. Are the needles sharp with four sides?	This is a spruce.	Go to 6.
5. Are the needles attached to the stem by a short green stalk?	Go to 6.	Go to 7.
6. Do the needles have gray stripes on the bottom?	This is a fir.	Go to 7.
7. Does the twig have one kind of leaf that looks like overlapping scales?	This is a cedar.	Go to 8.
8. Are the needles soft and flat?	This is a Douglas fir.	If you have not identified the conifer, return to 1.

Analysis and Conclusions

1. How did the dichotomous key help you to identify your conifer specimens?

    By answering a series of questions about the appearance of the conifer's needles, it is possible

    to identify each conifer.

2. Which kind of conifer has soft, flat needles?

    Douglas fir

3. Which kind of conifer has overlapping scales?

    cedar

Extension

Conifers can also be identified by the size and shape of their cones. Collect seed cones from at least five different conifers. Develop a dichotomous key based on the size and shape of the seed cones to identify each conifer.

Name_____ Class_____ Date_____

Chapter 10   Seed Plants

Laboratory 10-2 How are xylem and phloem arranged in woody and herbaceous stems?

Background Information

Xylem and phloem form a continuous transportation network in vascular plants. Xylem carries water up from the roots, through the stem, to the leaves. Phloem carries food down from the leaves to the stem, roots, and all other parts of the plant. Both herbaceous stems and woody stems contain xylem and phloem.

Skills: observing, comparing, inferring

Objectives

In this laboratory, you will
- observe and compare the location of xylem and phloem in woody and herbaceous stems.
- observe water transport in a stem.

Prelab Preparation

1. Review Section 10-1 Overview of Seed Plants and Section 10-3 Plant Tissues and Organs.

2. What is the difference between a herbaceous stem and a woody stem?

   A herbaceous stem is soft and green; a woody stem is not green and is harder and thicker than a herbaceous stem.

Materials

prepared slides of herbaceous and woody stems   compound microscope
2 beakers   food coloring
razor   metric ruler
2 stalks of celery with leaves

Procedure

1. Place a prepared slide of a herbaceous stem on the stage of a microscope.

2. Use low power to focus on the cross section of the stem. Then switch to high power.

3. Compare what you see with the cross sections of herbaceous stems shown in Figure 1. Locate the xylem and phloem.

4. Draw what you see in Plate 1. Label the xylem and phloem.

5. Place a prepared slide of a woody stem on the stage of a microscope.

6. Use low power to focus on the cross section of the stem. Then switch to high power.

7. Compare what you see with the cross section of a woody stem shown in Figure 2. Locate the xylem and phloem.

Biology   Copyright ©   by Globe Book Company

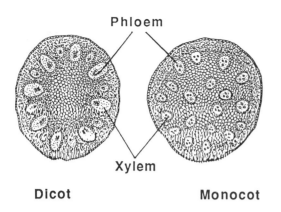

Figure 1: Herbaceous Stems

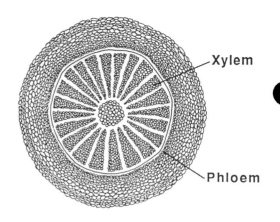

Figure 2: Woody Stem

8. Draw what you see in Plate 2. Label the xylem and phloem.

9. Fill two beakers half full of water. Add a few drops of food coloring to each beaker.

10. Use the razor to trim about 1 cm off the bottom of the celery stalk. Cut the leaves off of one stalk. **Caution: Be careful when using a razor.**

11. Place one stalk of celery into each beaker. Leave the stalks in the beaker for 20 minutes.

12. After 20 minutes, remove both stalks from the beakers.

13. Carefully cut cross sections of the stalk without leaves. Begin at the bottom and make a cut every 2 cm until food coloring is no longer visible in the stem.

14. Record the height to which the food coloring traveled up the stem in Data Table 1.

15. Repeat steps 13 and 14 for the stem with leaves.

Observations and Data

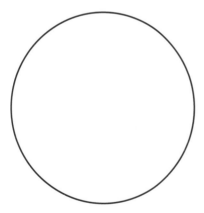

Plate 1: Herbaceous Stem

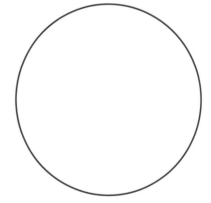

Plate 2: Woody Stem

Was your prepared slide of a herbaceous stem a monocot or a dicot? How do you know?

 Answers will vary. Students should compare their slides with Figure 1 to identify the stem.

Name _____ Class _____ Date _____

Chapter 10 Seed Plants

Laboratory 10-2

Data Table 1: Water Movement in Celery

Celery Stalk	Height of Coloring (cm)
Without leaves	**Answers will vary.**
With leaves	

In which celery stalk did the colored water climb higher — the stalk with leaves or the stalk without leaves?

the stalk with leaves

Analysis and Conclusions

1. The springwood produced by a woody stem is usually made up of large xylem tubes, while the summerwood is made up of thinner xylem tubes. How can you explain this difference?

   Answers will vary. In the spring, more moisture is available and more growth takes place than in the summer.

2. Is celery a herbaceous stem or a woody stem? How do you know?

   herbaceous, because it is soft and green

3. How can you explain the difference in height to which the colored water climbed in each celery stalk?

   Answers will vary. Because xylem transports water to the leaves, the water will not rise as fast if the leaves are missing.

Biology   Copyright © by Globe Book Company   49

Extension

Obtain a white carnation on its stem. Cut the stem so that it is 15 cm long. Carefully cut the stem in half lengthwise. Put one half of the stem into a beaker containing a solution of red food coloring. Put the other half into a beaker containing a solution of blue food coloring. Leave the carnation and beakers undisturbed for 24 hours. Predict what you think will happen. Observe the carnation after 24 hours. Was your prediction correct? Explain. Use the space provided to record your prediction and observations.

Name _____ Class _____ Date _____

Chapter 11   Plant Structure and Function

Laboratory 11-1 How do different conditions affect the rate of germination of lima bean seeds?

Background Information

A seed is the dormant, or resting, stage of a plant. A seed can remain dormant when conditions of moisture, light, or temperature are not suitable for growth. When conditions are favorable, the seed will begin to grow, or germinate. Some seeds will germinate only under special conditions, for example, being exposed to freezing temperatures or being acted on by the digestive juices of animals.

Skills: observing, measuring, recording data

Objective

In this laboratory, you will
- observe how different conditions affect the rate of germination of lima bean seeds.

Prelab Preparation

1. Review the procedure for germinating seeds.
2. Review how to find an average.

Note to the teacher: You may wish to soak the lima beans in advance and make them available to the students on the day of the laboratory.

Materials

lima beans	vinegar	water
paper towels	6 baby food jars	metric ruler
hand lens	marking pencil	

Procedure

1. Soak four lima beans overnight in vinegar. Soak 20 lima beans overnight in water.

2. Label six baby food jars 1 through 6.

3. Fold six paper towels and place one towel in each jar. Moisten each towel with water.

4. Place the four lima beans that were soaked in vinegar into jar 1. Be sure the seeds are between the moist paper towel and the side of the jar.

5. Place four lima beans that were soaked in water into each of the remaining jars (jars 2–6).

6. Put jars 1 and 2 in a dark location where they will not be disturbed.

7. Put jar 3 on a windowsill where it will receive sunlight for several hours each day. Put jar 4 in a dark location.

8. Put jar 5 in a refrigerator. Put jar 6 in a warm, dark location.

9. Observe each jar every day for five days. Measure the amount of new growth on each bean seed and calculate the average new growth for the four seeds in each jar.

10. Record your observations in Data Table 1.

Biology   Copyright ©   by Globe Book Company

Observations and Data

Data Table 1: Seed Growth

Jar	Average New Growth (mm)				
	Day 1	Day 2	Day 3	Day 4	Day 5
1 (vinegar)			Answers will vary.		
2 (water)					
3 (light)					
4 (dark)					
5 (cold)					
6 (warm)					

Analysis and Conclusions

1. What three variables were tested in this laboratory?

   presence of acid, light, temperature

2. Which conditions were most favorable for the growth of lima bean seeds? Use your data to support your answer.

   Answers will vary. Beans soaked in water and placed in a warm, dark location will germinate faster than those soaked in vinegar, placed in sunlight, or kept in the refrigerator.

Extension

How do pollutants affect the germination of lima bean seeds? Design an experiment to answer this question. Soak lima beans in water, in a detergent solution, in liquid fertilizer, and in other solutions of common household chemicals. Compare the rate of germination of the seeds over a five-day period. Record your observations in a data table.

Name _____ Class _____ Date _____

Chapter 11   Plant Structure and Function

Laboratory 11-2 How do plants reproduce by vegetative propagation?

Background Information

Plants can reproduce sexually or asexually. Asexual reproduction in plants is called vegetative propagation. One method of vegetative propagation involves the use of cuttings. A plant cutting, usually a stem or leaf, is placed in water until roots form. The cutting is then planted in soil and a new plant begins to grow. Plants such as onions grow from underground bulbs. New plants can be grown from the bulbs. Have you ever seen a potato with "eyes"? Each potato "eye" is a bud that can grow into a new plant.

Skills: observing, comparing, inferring

Objective

In this laboratory, you will
- observe the growth of plants using three methods of vegetative propagation.

Prelab Preparation

Compare the following terms:
1. sexual reproduction, asexual reproduction

   Sexual reproduction involves two parents; asexual reproduction involves only one parent.

2. seed, bud

   Both a seed and a bud contain embryo plants and will grow into new plants; a seed is a result of sexual reproduction, while a bud is a form of asexual reproduction.

Materials

3 clear plastic cups	toothpicks	water
begonia plant	white potato	onion
razor		

Procedure

1. Use a razor to cut a piece of stem from a begonia plant. **Caution: Be careful when using a razor.** Be sure to cut below a leaf so that several leaves remain on the cut stem.

2. Fill a plastic cup with water. Place the begonia cutting in the cup. See Figure 1.

3. Put the cup with the cutting in a location where it will not receive direct sunlight and where it will not be disturbed.

4. Observe your cutting every day for five days. Notice when roots first begin to develop. Record your observations in Data Table 1.

5. Insert four toothpicks around the center of an onion.

6. Fill a plastic cup with water. Use the toothpicks to support the onion in the cup of water, as shown in Figure 2.

7. Put the cup with the onion in a location where it will not receive direct sunlight and where it will not be disturbed.

8. Observe the onion every day for five days. Notice when roots and leaves begin to develop. Record your observations in Data Table 1.

Biology   Copyright ©   by Globe Book Company   53

9. Repeat steps 5 to 8 with the potato. See Figure 3.

Figure 1: Cutting

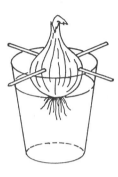

Figure 2: Bulb

Figure 3: Bud

Observations and Data

Data Table 1: Vegetative Propagation

Sample	Growth				
	Day 1	Day 2	Day 3	Day 4	Day 5
Begonia cutting			Answers will vary.		
Onion bulb					
Potato "eyes"					

1. Which method of vegetative propagation was the first to show new growth?
   Answers will vary.

2. Which method was the slowest in developing new growth?
   Answers will vary.

Analysis and Conclusions

1. Compare your results with those of your classmates. Did everyone obtain the same results?
   Answers will vary.

2. What conditions do you think could have caused any different results in your class?
   Answers will vary. Possible conditions include the following: amount of water; temperature; amount of sunlight; damage to cuttings; differences among plants, onions, and potatoes, and so forth.

Extension

Obtain several clay pots, gravel, and potting soil. Plant each of your samples in a separate pot. Water each pot carefully. Place the pots in a location where they will receive light. Observe each pot for the growth of new plants. Try growing other plants by vegetative propagation. Choose the correct method for the following plants: a carrot, turnip, African violet, geranium, coleus, tulip, and garlic.

Name _____ Class _____ Date _____

Chapter 11   Plant Structure and Function

Laboratory 11-3   What are the parts of a perfect and an imperfect flower?

Background Information

Flowers are the reproductive structures of flowering plants. The male parts of a flower are called stamens; the female parts are called pistils. Some flowers have both stamens and pistils. These flowers are called perfect flowers. Flowers that have only stamens or only pistils are imperfect flowers. Flowers with only stamens are staminate; those with only pistils are pistillate. The delicate stamens and pistils are surrounded and protected by the petals. Monocot flowers have petals in multiples of three. Dicot flowers have petals in multiples of four or five.

Skills: observing, comparing, analyzing

Objective

In this laboratory, you will
- examine and compare the parts of three different flowers.

Prelab Preparation

1. Review Section 11-3 Flowers.

2. What are the two parts of a stamen?

   anther, filament

3. What are the three parts of a pistil?

   stigma, style, ovary

Materials

rose	begonia	tulip
scalpel	forceps	hand lens

Procedure

1. Compare each of your flowers with the typical flower shown in Figure 1.

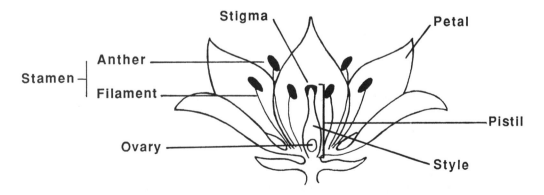

Figure 1: Parts of a typical flower

Biology   Copyright ©   by Globe Book Company   55

2. Gently remove the petals from the rose. Count the number of petals and record the number of petals in Data Table 1.

3. Repeat step 2 with the tulip and the begonia.

4. Find the reproductive parts of each flower. Carefully remove each reproductive part by cutting it off at the base with a scalpel. **Caution: Be careful not to cut yourself when using a scalpel.**

5. Examine each reproductive part carefully with a hand lens.

6. Draw and label the parts of each flower in the space provided.

7. Complete Data Table 1. If a flower has the part listed, place a check mark in the correct column.

Observations and Data

Data Table 1: Parts of Flowers

Parts	Flowers		
	Rose	Tulip	Begonia
Number of Petals	5 or 10	3 or 6	3 or 6
Stamens	√	√	Answers will vary.
Pistils	√	√	

Rose	Tulip	Begonia
	Students should draw a stamen and a pistil for perfect flowers and either a stamen or a pistil for imperfect flowers. Labels should include anther, filament, stigma, style, and ovary.	

Name _____ Class _____ Date _____

Chapter 11 Plant Structure and Function   Laboratory 11-3

Analysis and Conclusions

1. a. Which flowers are monocots? How do you know?

   tulip, begonia

   They have petals in multiples of 3.

   b. Which are dicots? How do you know?

   rose

   It has petals in multiples of 5.

2. a. Which flowers are perfect?

   rose, tulip

   b. Which are imperfect?

   begonia

   c. Are the imperfect flowers staminate or pistillate?

   Answers will vary. Staminate and pistillate flowers often occur

   in the same cluster in begonias.

3. What is the difference between a perfect flower and an imperfect flower?

   A perfect flower has both stamens and pistils. An imperfect flower has

   either stamens or pistils, but not both.

Biology   Copyright © by Globe Book Company   57

Extensions

1. Obtain a few pollen grains from the anther of a stamen. Make a wet-mount slide of the pollen grains. Use the low power of a compound microscope to examine the pollen grains. Sketch the magnified pollen grains in the space provided.

2. With a scalpel, carefully cut a pistil in half lengthwise. **Caution: Be careful not to cut yourself when using a scalpel.** Examine the ovules in the ovary with a hand lens. Sketch the ovules and ovary in the space provided.

Name _____ Class _____ Date _____

Chapter 12　　　　　　　　　　　　　　　　　　　Porifera and Cnidarians

Laboratory 12-1　How does a hydra respond to stimuli?

Background Information

All animals react to sudden changes in the environment. A change in the environment is a stimulus; the animal's reaction is a response. Different animals may respond differently to the same stimulus. An animal that can move freely may move toward or away from a stimulus.

Skills: observing, predicting, inferring

Objective

In this laboratory, you will
- observe the response of hydra to two different stimuli.

Prelab Preparation

Review Section 12-2 Cnidarians.

Materials

hydra culture　　　　　　　　compound microscope　　　　　　flashlight
depression slides　　　　　　cover slips　　　　　　　　　　　dropper
toothpick　　　　　　　　　　vinegar

Procedure

1. Using a dropper, place a sample of the hydra culture on a clean, dry depression slide. Place a cover slip over the sample.

2. Place the slide on the stage of a compound microscope. Observe the hydra with the low power of the microscope. Draw and label the parts of a hydra in Plate 1.

3. Predict how you think the hydra will respond to a beam of light. Record your prediction in Data Table 1.

4. Use a flashlight to shine a beam of light onto the slide of the hydra sample. Observe the response of the hydra to the light. Record your observations in Data Table 1.

5. Predict how you think the hydra will respond to the presence of vinegar. Record your prediction in Data Table 1.

6. Dip a toothpick in vinegar. Place a drop of vinegar at the edge of the cover slip.

7. Observe the response of the hydra to the vinegar. Record your observations in Data Table 1.

Biology　　　　　　　　　　Copyright ©　　by Globe Book Company　　　　　　　　　　59

Observations and Data

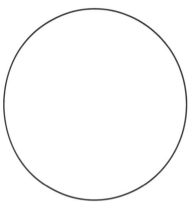

Plate 1: Hydra

Data Table 1: Response of Hydra to Stimuli

Stimulus	Response	
	Predicted	Actual
Light	Answers will vary.	moved away
Vinegar		released nematocysts

Were your predictions correct?

Answers will vary.

Analysis and Conclusions

1. Describe the response of hydra to (a) light and (b) vinegar.

   (a) The hydra contracted or moved away from the light.

   (b) The hydra released their nematocysts.

2. Did the hydra respond differently to light and to vinegar? Why?

   Answers will vary. Accept all logical responses.

Extension

Design an experiment to find out how hydra respond to changes in temperature and to different concentrations of salt in their environment. Record your predictions and the actual responses of the hydra in a data table.

Name _____ Class _____ Date _____

Chapter 12   Porifera and Cnidarians

Laboratory 12-2   What do the support structures of sponges look like?

Background Information

Sponges do not have a skeleton for support. Instead, some sponges have a network of protein fibers called spongin. Other sponges have tiny, hard particles called spicules. Spicules are made up of calcium carbonate or silicon dioxide.

Skills: observing, analyzing, comparing

Objective

In this laboratory, you will
- observe the support structures of two sponges.

Prelab Preparation

Review how to make a wet-mount slide.

Materials

compound microscope	2 toothpicks	dropper
2 microscope slides	small beaker	bleach
2 cover slips	pieces of *Grantia* and *Spongia*	

Procedure

1. Pour 10 mL of bleach into a small beaker. **Caution: Bleach is a caustic base. If you spill any bleach on your skin or clothes, rinse with plenty of water.**

2. Use scissors to cut off a small piece of *Grantia*. **Caution: Be careful when using scissors.**

3. Place the piece of *Grantia* on a microscope slide.

4. With a dropper, place two or three drops of bleach on the piece of *Grantia*. Gently mix the bleach into the *Grantia* with a toothpick.

5. Lower a cover slip onto the slide to make a wet mount. Observe the *Grantia* using the low power of a microscope.

6. Draw what you observe in Plate 1.

7. Repeat steps 2–5 with the *Spongia*. Draw what you observe in Plate 2.

8. Clean your work area. Follow your teacher's instructions for proper disposal of the remaining sponge pieces and bleach.

9. Wash your hands with soap and water.

Biology   Copyright ©   by Globe Book Company   61

Observations and Data

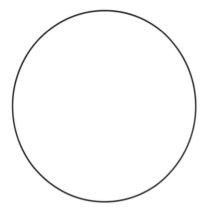

Plate 1: *Grantia*

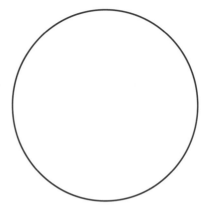

Plate 2: *Spongia*

Analysis and Conclusions

1. Why did you add bleach to your sponge pieces?

   The bleach dissolved the cellular material of the sponge.

2. a. Which sponge contained spicules?

   Grantia

   b. Which sponge contained spongin?

   Spongia

3. Describe the appearance of the spicules.

   Answers will vary. Spicules should appear glassy and look like spikes or needles.

4. Describe the appearance of the spongin.

   Answers will vary. Spongin should look like fibers.

5. Which sponge would you most likely use to wash a car or window? Explain you answer.

   Spongia, because the needlelike spicules of Grantia might scratch the car or window surface.

Extension

Obtain a natural sponge such as *Grantia*. Observe the sponge with a hand lens. Draw a sketch of the sponge and label the pores and osculum. Carefully cut the sponge in half lengthwise. **Caution: Be careful when using a scissors.** Draw a diagram of the inside of the sponge. Use arrows to show how water enters and leaves the sponge.

Name _____ Class _____ Date _____

Chapter 13 | Worms

Laboratory 13-1 What are the external and internal structures of an earthworm?

Background Information

The organ systems of earthworms are more highly developed than those of nonsegmented worms. The organs develop in a central body cavity and are covered by a thin membrane that holds them in place. An earthworm has a digestive system, a nervous system, a circulatory system, a reproductive system, and a muscular system.

Skills: observing, comparing, analyzing

Objective

In this laboratory, you will
- observe the external and internal structures of an earthworm.

Prelab Preparation

1. Review Section 13-4 Segmented Worms.

2. a. What does the term "annelid" mean?

   "little rings"

   Note to the teacher: This laboratory exercise is optional. You may wish to have students identify the structures listed using models, overhead transparencies, blackline masters, computer simulations, and so forth.

   b. To what structures of an earthworm does this term apply?

   to the body segments

Materials

preserved earthworm
dissecting needle
hand lens

dissecting pan
scalpel
forceps

dissecting pins
scissors

Procedure

1. **Caution: Wear gloves during this laboratory.** Rinse a preserved earthworm with running water to remove as much preservative as possible.

2. Place the earthworm in a dissecting pan. Locate the anterior and posterior ends of the earthworm.

3. Rub your fingers over the bottom (ventral) surface of the worm to feel the setae.

4. Refer to Figure 1. Locate the following structures: prostomium, mouth, body segments, clitellum, setae, anus, seminal receptacle openings, sperm duct openings, oviduct openings.

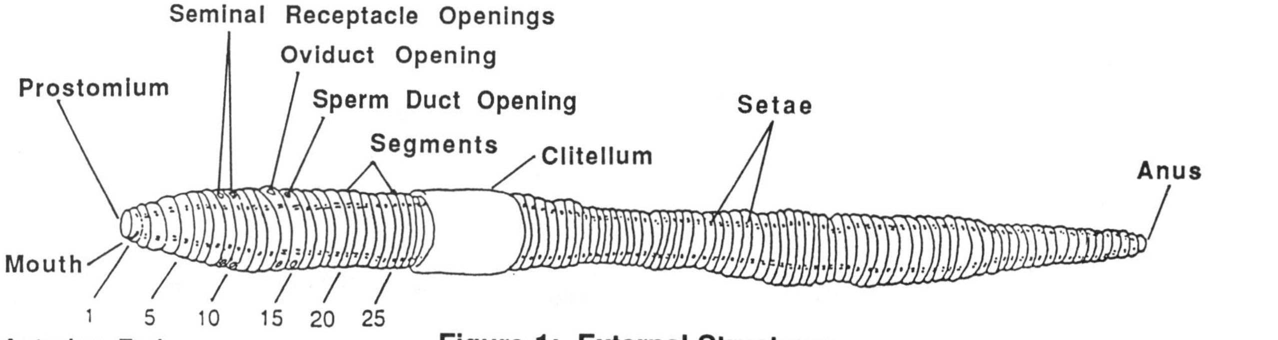

Figure 1: External Structures

Biology — Copyright © by Globe Book Company

5. Turn the dissecting pan so that the anterior end of the earthworm is facing away from you.

6. Stretch the worm, dorsal side up, and place pins through the first and last segments to hold it in place.

7. **Caution: Be careful when using a scalpel.** Beginning behind the clitellum, make a cut through the body wall slightly to one side of the midline. Be sure to cut only the body wall so as not to damage the internal organs. Continue the cut to the anus.

8. Turn the pan so that the anterior end of the earthworm is facing toward you. Extend the cut to the mouth.

9. Hold the body wall back with forceps. Use the scalpel to cut through the membranes that separate the segments of the earthworm. Refer to Figure 2.

10. After every 15 segments, fold the body wall back and place a pin at a 45-degree angle to hold the body wall in place. Refer to Figure 3.

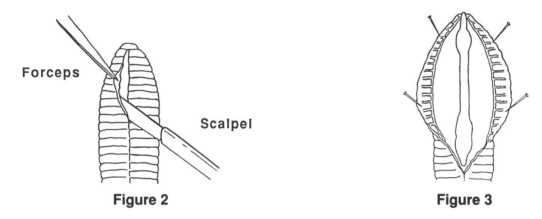

Figure 2   Figure 3

11. Refer to Figure 4 to locate the following organs of the digestive system: pharynx, esophagus, crop, gizzard, intestine.

12. Locate the following organs of the circulatory system: dorsal blood vessel, aortic arches, ventral blood vessel.

13. Locate the following organs of the nervous system: "brain," ventral nerve cord.

14. Locate the following organs of the reproductive system: seminal vesicles, seminal receptacles, testes, ovaries. Use a hand lens if necessary.

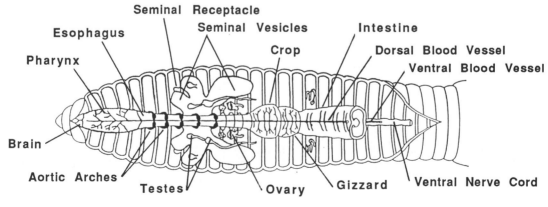

Figure 4: Internal Organs

Name _____ Class _____ Date _____

Chapter 13   Worms

Laboratory 13-1

Observations and Data

1. Describe the prostomium.

   The prostomium is a liplike structure above the mouth at the anterior end of the earthworm.

2. Describe the clitellum.

   The clitellum is a thickened area circling the earthworm's body near the anterior end.

Analysis and Conclusions

1. Does the earthworm have an open or a closed circulatory system? How do you know?

   closed, because the blood is carried in tubes, or vessels

2. Why do you think the body structure of an earthworm is often called "a tube within a tube"?

   The outer body resembles a tube; the internal organs are contained within a tube, or body cavity.

3. Is the earthworm male or female? How do you know?

   An earthworm is both male and female, because each individual has both male and female reproductive

   organs.

Biology   Copyright ©   by Globe Book Company

Extension

Use reference materials to find out how leeches, which are segmented worms, were used in medicine in the past and how they are being used today. Use the space provided to record your findings. Include a diagram of a leech.

Name_____ Class_____ Date_____

Chapter 13   Worms

Laboratory 13-2 How does an earthworm respond to stimuli?

Background Information

Like all animals, earthworms respond to stimuli, or changes in their environment. Earthworms receive and react to information from the environment. The information may be in the form of physical or chemical stimuli such as odors, sounds, tastes, touch, and light. The responses of an earthworm to these stimuli determine its behavior.

Skills: predicting, observing, analyzing

Objectives

In this laboratory, you will
- predict how you think an earthworm will respond to various stimuli.
- observe the responses of an earthworm to various stimuli.

Prelab Preparation

1. Review Section 13-4 Segmented Worms.

2. Why must an earthworm's skin be kept moist?

   <u>The earthworm does not have a respiratory system. It takes in oxygen and gives off carbon dioxide directly</u>

   <u>through its skin.</u>

   _____

Materials

live earthworm   dissecting pan   paper towels
water   dropper   cotton swabs
vinegar   toothpick   flashlight
ground meat   black paper

Procedure

1. Line a dissecting pan with paper towels. Moisten the paper towels with water. **Caution: Be sure to keep the earthworm's skin moist at all times during this laboratory.** Place an earthworm into the dissecting pan. Observe the earthworm for several minutes. Notice its movements and reactions.

2. Predict how you think the earthworm will react to a wet cotton swab held about 3 cm from its anterior end. Record you prediction in Data Table 1.

3. Dip a cotton swab into water and hold it about 3 cm from the anterior end of the earthworm. Observe the earthworm's response. Record your observation in Data Table 1.

4. Predict how you think the earthworm will react to a cotton swab soaked in vinegar. Record your prediction in Data Table 1.

5. Dip a cotton swab into vinegar and hold it about 3 cm from the anterior end of the earthworm. Record the earthworm's response in Data Table 1.

Biology   Copyright ©   by Globe Book Company

6. Predict how you think the earthworm will respond to touch. Record your prediction in Data Table 1.

7. With a toothpick, gently touch the earthworm near the anterior end, posterior end, and clitellum. Record the earthworm's response in Data Table 1.

8. Predict how you think the earthworm will react to a beam of light on its anterior end. Record your prediction in Data Table 1.

9. Cover the posterior end of the earthworm with a piece of black paper. Shine the light from a flashlight on the anterior end. Record the earthworm's response in Data Table 1.

10. Predict how you think the earthworm will respond to a beam of light on its posterior end. Record your prediction in Data Table 1.

11. Cover the anterior end of the earthworm with a piece of black paper. Shine the light from a flashlight on the posterior end. Record the earthworm's response in Data Table 1.

12. Predict how you think the earthworm will respond to ground meat. Record your prediction in Data Table 1.

13. Place a small piece of ground meat about 3 cm from the anterior end of the earthworm. Record the earthworm's response in Data Table 1.

14. Return the earthworm to its natural environment. Follow your teacher's instructions.

Observations and Data

Data Table 1: Stimuli and Responses

Stimulus	Response	
	Predicted	Actual
Water	Answers will vary.	moves toward
Vinegar		moves away
Touch		moves away
Light: anterior end		moves away
posterior end		none
Ground meat		moves toward

Name_____ Class_____ Date_____

Chapter 13 Worms   Laboratory 13-2

Analysis and Conclusions

1. Why was it necessary to observe the earthworm for several minutes before testing its responses to stimuli?

   Answers will vary. Observing the earthworm's "normal" behavior provides a control, or basis for comparison with its behavior in response to stimuli.

2. Why did the earthworm respond differently to light on its anterior end and on its posterior end?

   Answers will vary. Students should indicate that the earthworm's "brain" is located in the anterior end.

3. Why do you think the earthworm's response to light is important to its survival?

   Answers will vary. Avoiding direct light helps keep the earthworm's skin from drying out.

Extension

After a rainfall, earthworms can usually be found on sidewalks and roads. What do you think causes this behavior? Design an experiment to test an earthworm's response to varying amounts of water. Be sure to follow proper procedures for the care and handling of live animals in the laboratory. Use the space provided on the following page to describe your experiment. Record your observations in a data table.

Biology   Copyright ©   by Globe Book Company

NOTES

Name _____ Class _____ Date _____

Chapter 14   Mollusks and Echinoderms

Laboratory 14-1 What are the external and internal structures of a sea star?

Background Information

Sea stars are members of the phylum *Echinodermata*, the "spiny-skinned" animals. Sea stars have radial symmetry; that is, they have no anterior or posterior end. The parts of a sea star radiate out from the center, like spokes from the hub of a wheel.

Skills: observing, comparing, analyzing

Objective

In this laboratory, you will
- observe the external and internal structures of a sea star.

Note to teacher: This laboratory exercise is optional. You may wish to have students identify the structures listed using models, overhead transparencies, blackline masters, computer simulations, and so forth.

Prelab Preparation

1. Review Section 14-2 Echinoderms.

2. What are the dorsal and ventral surfaces of a sea star?

   The dorsal surface is the top side; the ventral surface is the bottom side.

Materials

preserved sea star	dissecting pan	forceps
scissors	hand lens	water
dropper		

Procedure

1. **Caution: Wear gloves during this laboratory.** Rinse a preserved sea star under running water to remove as much preservative as possible.

2. Place the sea star into a dissecting pan, dorsal surface up. Locate the central disk surrounded by five rays, or arms.

3. Find the sieve plate. Observe this structure with a hand lens.

4. Use the hand lens to observe the spines that cover the surface of the sea star.

5. Locate an eyespot at the tip of an arm and observe it with the hand lens.

6. Turn the sea star over so that the ventral surface is facing up.

7. Locate the mouth opening in the middle of the central disk.

8. Find and examine the grooves extending from the mouth to the tip of each arm.

9. Use the hand lens to examine the tube feet lining each side of a groove.

10. Label the following structures on Figure 1: central disk, arm, sieve plate, spines, eyespot, mouth, groove, tube feet.

Biology   Copyright © by Globe Book Company   71

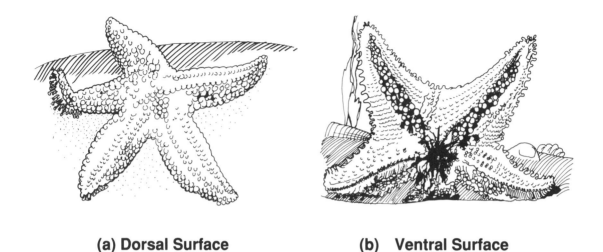

(a) Dorsal Surface   **(b) Ventral Surface**

Figure 1: Sea Star

11. Turn the sea star over so that the dorsal surface is up.

12. Using the scissors, carefully cut the tips off two arms. **Caution: Be careful when using a scissors.**

13. Use the scissors to make cuts 1, 2, and 3 as shown in Figure 2.

14. Carefully remove the flap of skin so that you can observe the internal structures of the sea star. If necessary, add a few drops of water to the exposed parts to keep them from drying out.

15. Remove the long, greenish digestive glands from the arm.

16. Observe the reproductive structures under the digestive glands. Ovaries should appear orange; testes should appear gray.

17. Remove part of the central disk. Locate the stomach under the central disk. Find the tube leading to the stomach from the digestive gland.

18. Using the hand lens, examine the cut end of the second arm. Locate and examine the tube feet.

19. Complete Data Table 1.

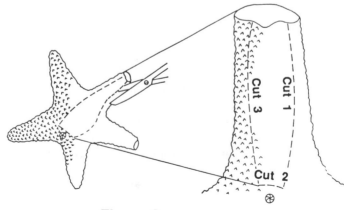

Figure 2

Name _____ Class _____ Date _____

Chapter 14 Mollusks and Echinoderms   Laboratory 14-1

Observations and Data

Data Table 1: Organs of a Sea Star

System	Organs	Location
Digestive	mouth	central disk (ventral surface)
	stomach	under central disk
	digestive glands	arms
Reproductive	ovaries	arms
	testes	arms
Nervous	eyespots	tips of arms
Water vascular	sieve plate	central disk (dorsal surface)
	tube feet	arms (ventral surface)

1. Does the sea star have an exoskeleton or an endoskeleton?
   an endoskeleton covered by a spiny skin

2. Is your sea star male or female? How do you know?
   Answers will vary, depending on the color of the gonads. Note to teacher: The reproductive structures in a preserved sea star may lose their color.

Analysis and Conclusions

1. What are some possible advantages of radial symmetry for a sea star?
   Accept all logical answers.

2. The diet of sea stars consists mainly of mollusks. What are two adaptations to this diet that you observed in your sea star?
   Answers will vary. Students should mention tube feet (to open a mollusk shell) and a stomach that can be inverted to digest a mollusk in its shell.

Biology   Copyright © by Globe Book Company   73

Extension

If possible, obtain a preserved sea cucumber and observe its external and internal structures. Compare the structure of a sea cucumber with the structure of a sea star. If preserved specimens are not available, use reference materials to research sea cucumbers and other echinoderms. Use the space provided to list the similarities and differences in their structures and functions.

Name _____ Class _____ Date _____

Chapter 15 — Arthropods

Laboratory 15-1 How do the external features of a crayfish help to identify it as an arthropod?

Background Information

Crayfish belong to a class of arthropods called crustaceans. Like all arthropods, crustaceans have jointed legs, an exoskeleton, and a body divided into segments. The exoskeleton is made up mostly of a nonliving material called chitin.

Skills: observing, classifying, analyzing

Objective

In this laboratory, you will
- observe the external features of a crayfish that identify it as an arthopod.

Prelab Preparation

Note to teacher: This laboratory exercise is optional. You may wish to have students identify the structures listed using models, overhead transparencies, blackline masters, computer simulations, and so forth.

1. Review Section 15-2 Crustaceans.

2. How are crustaceans similar to arachnids?
   They both have two main body parts.

3. How are arachnids and crustaceans different from insects?
   Insects have three main body parts.

4. What is a carapace?
   the part of the exoskeleton that covers the head-chest region

Materials

preserved crayfish   dissecting pan
dissecting needle   hand lens

Procedure

1. **Caution: Wear gloves during this laboratory.** Rinse a preserved crayfish under running water to remove as much preservative as possible.

2. Place the crayfish into a dissecting pan. Observe the two body segments: head–chest and abdomen.

3. Locate the carapace and observe that it covers the upper surface and sides of the head–chest region.

4. Observe the anterior end of the crayfish. Locate the eyes at the ends of movable eye stalks. Use the hand lens to examine the structure of the eyes.

5. Locate the long antennae and the shorter antennules.

6. Locate the five pairs of legs attached to the head–chest region. The first pair, called chelipeds, end in large pincers. The other four pairs are walking legs.

7. Observe the abdomen. Locate the five pairs of jointed appendages called swimmerets.

8. Use the hand lens to observe the first two pairs of swimmerets. In males, the first two pairs are larger than the other swimmerets. In females, the first two pairs are smaller than the other swimmerets.

9. Observe the last segment of the abdomen. Locate a pair of fan-shaped structures called uropods and a triangular structure called the telson.

10. Label the following features on Figure 1: carapace, abdomen, eye, eye stalk, antenna, antennules, cheliped, walking legs, swimmerets, uropod, telson.

Observations and Data

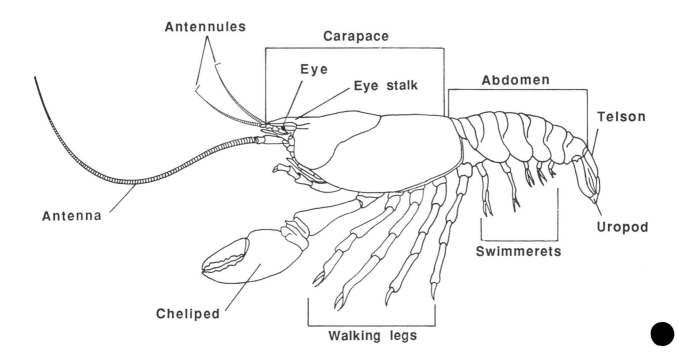

Figure 1: External Structures

1. Which of the walking legs have claws?
   the first and second pairs

2. Is your crayfish male or female? How do you know?
   Answers will vary depending on the appearance of the first two pairs of swimmerets.

Analysis and Conclusions

1. What is a possible advantage of having eyes on movable stalks?
   Eyes on movable stalks allow a crayfish to see more of its surroundings than would eyes without stalks.

2. What three features of a crayfish identify it as an arthropod?
   1. jointed legs   2. exoskeleton   3. body divided into segments

Extension

Isopods, such as sow bugs and pill bugs, are also crustaceans. They live in damp areas on land. Obtain several live isopods. Observe the responses of the isopods to various stimuli, such as light and moisture. Be sure to follow proper procedures for handling live animals in the laboratory. Record your observations in a data table.

Name_____ Class_____ Date_____

Chapter 15   Arthropods

Laboratory 15-2 How do the external features of a grasshopper help to identify it as an arthropod?

Background Information

The arthropod phylum includes more species of animals than all the other phyla of the animal kingdom combined. Arthropods are characterized by jointed legs, an exoskeleton, and a body divided into segments. Insects, such as grasshoppers, are arthropods that make up the largest and most successful group of organisms alive today.

Skills: observing, comparing, analyzing

Objective

In this laboratory, you will
- observe the external features of a grasshopper that identify it as an arthropod.

Prelab Preparation

Note to teacher: This laboratory exercise is optional. You may wish to have students identify the structures listed using models, overhead transparencies, blackline masters, computer simulations, and so forth.

1. Review Section 15-5 Insects.

2. What is the difference between an endoskeleton and an exoskeleton?

   An endoskeleton is internal and is made up of living bone tissue.

   An exoskeleton is external and is made up of nonliving material.

Materials

preserved grasshopper   dissecting pan
dissecting needle   hand lens

Procedure

1. **Caution: Wear gloves during this laboratory.** Rinse a preserved grasshopper under running water to remove as much preservative as possible.

2. Place the grasshopper into a dissecting pan. Observe that the grasshopper is divided into three segments: head, thorax, and abdomen.

3. Examine the grasshopper's head. Locate the compound eyes, antennae, and mouth parts. Examine the compound eyes with a hand lens.

4. Use the hand lens to locate the simple eyes. The simple eyes are arranged in a triangle on the front of the head. Two simple eyes are located in front of the compound eyes; the third is between the antennae.

5. Examine the three pairs of legs that are attached to the thorax. Notice the hooks on the last segment of each leg.

6. Observe and compare the two pairs of wings. Lift the wings and use the hand lens to locate the tympanum.

7. Use the hand lens to locate the spiracles on each side of the abdomen.

8. Observe the tip of the abdomen. The abdomen is tapered and ends in a structure called the ovipositor. In females, the ovipositor is hard and pointed.

Biology   Copyright ©   by Globe Book Company

9. Label the following features on Figure 1: head, thorax, abdomen, antennae, simple eyes, compound eyes, mouth, forewing, hindwing, spiracles, ovipositor.

Observations and Data

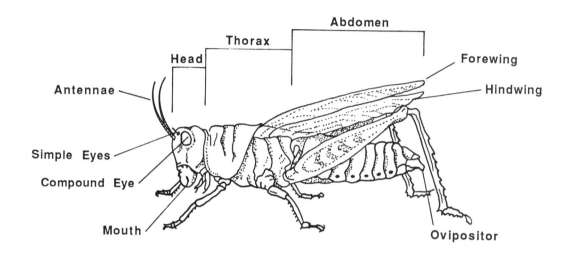

Figure 1: External Structures

1. Compare the appearance of the forewings and hindwings.
   The forewings are tough and leatherlike; the hindwings are made of thin membranes.

2. Is your grasshopper male or female? How do you know?
   Answers will vary depending on the shape of the ovipositor.

Analysis and Conclusions

1. Which pair of legs is adapted for jumping? How do you know?
   the third pair, because they are longer and stronger than the two other pairs

2. What do you think is the function of the forewings and hindwings?
   Answers will vary. The forewings cover and protect the hindwings; the hindwings are used for flying.

3. What three features of a grasshopper identify it as an arthropod?
   1. jointed legs   2. exoskeleton   3. body divided into segments

Extension

Wasps are insects. Obtain a preserved specimen of a wasp or use a model. Observe the external features of the wasp. Draw and label a diagram showing the external features of a wasp. Compare your diagram of the wasp with the diagram of the grasshopper.

Name _____ Class _____ Date _____

Chapter 16 — Fishes

Laboratory 16-1 How do the external and internal structures of a perch adapt it for life in the water?

Background Information

All fishes belong to the phylum *Chordata*. Fishes are divided into three classes: jawless, cartilaginous, and bony fishes. The perch is an example of a bony fish. It is a ray–finned fish with a streamlined body and spiny fins that help it swim.

Skills: observing, comparing, analyzing

Objectives

In this laboratory, you will
- observe the external and internal structures of a perch.
- identify how the structures are adaptations to life in the water.

Note to teacher: This laboratory exercise is optional. You may wish to have students identify the structures listed using models, overhead transparencies, blackline masters, computer simulations, and so forth.

Prelab Preparation

1. Review Section 16-4 Bony Fishes.

2. What is the difference between a cartilaginous fish and a bony fish?

   A cartilaginous fish has a skeleton made up of cartilage, while a bony fish has a skeleton made up of bone.

Materials

preserved perch   dissecting pan   dissecting needle
scissors   forceps   hand lens

Procedure

1. **Caution: Wear gloves during this laboratory.** Rinse a preserved perch under running water to remove as much preservative as possible.

2. Place the perch into a dissecting pan as shown in Figure 1. Locate the head, trunk, and tail.

Figure 1

Biology   Copyright © by Globe Book Company   79

3. Examine the perch's head. Locate the eyes, mouth, and teeth. **Caution: Be careful when examining the teeth; they are very sharp.**

4. Locate the gill cover, or operculum. Lift the operculum and examine the gills beneath it. Notice the number and appearance of the gills.

5. Examine the fins of the perch. Identify the pectoral, dorsal, pelvic, anal, and caudal fins. Complete Data Table 1.

6. Locate the lateral line of the perch.

7. Use a hand lens to examine the scales.

8. Label the following structures on Figure 2: eye, mouth, operculum, pectoral fin, pelvic fin, dorsal fin, anal fin, caudal fin, lateral line, scales.

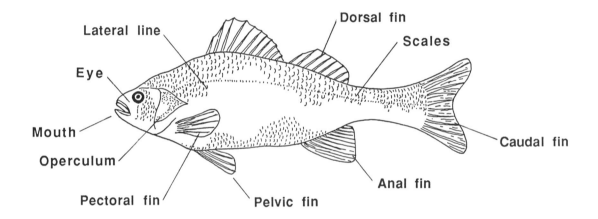

Figure 2

9. Turn the fish so that the ventral surface is facing up. Locate the anus. Carefully insert the point of the scissors into the body wall directly in front of the anus. Caution: Be careful when using a scissors. Do not cut deeply enough to damage the internal organs. Refer to Figure 3.

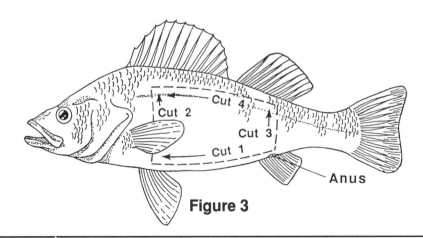

Figure 3

Name _____ Class _____ Date _____

Chapter 16   Fishes

Laboratory 16-1

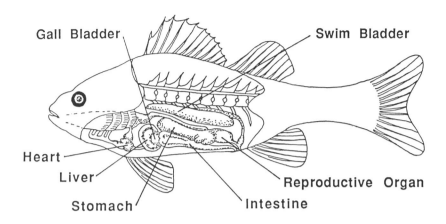

Figure 4

10. Carefully cut from the anus forward toward the pectoral fin.

11. Make another cut upward from the pectoral fin toward the dorsal fin. Make a third cut upward from the anus toward the dorsal fin. This will form a flap of skin.

12. Make a fourth cut to remove the flap of skin. This will expose the internal organs.

13. Refer to Figure 4. Locate the following organs of the digestive system: liver, gallbladder, stomach, and intestine. You may need to remove the liver to see the stomach.

14. Locate the swim bladder.

15. Locate the two–chambered heart in the cavity below the gills.

16. Locate the reproductive organs. The reproductive organs will vary depending on the sex of the perch and the time of year.

17. Complete Data Table 2.

Observations and Data

Data Table 1   Fins of a Perch

Fin	Number	Location
Pectoral	2	behind gills
Dorsal	2	dorsal surface
Pelvic	2	ventral surface in front of anal fin
Anal	1	ventral surface behind anus
Caudal	1	tail

Biology

Data Table 2: Internal Organs of a Perch

Organs	Description
Liver	large, cream colored
Gallbladder	small, round, below liver
Stomach	large, light-colored pouch
Intestine	long tube coming from stomach
Swim bladder	long, balloonlike
Heart	two-chambered, muscular, below gills
Reproductive organ	large, light colored, below swim bladder

Analysis and Conclusions

1. The perch is a carnivore. How are its jaw and teeth adapted for eating other fish?
   The perch has a large jaw for catching prey and sharp teeth for holding and tearing.

2. Which fin is adapted for moving the perch through the water?
   caudal

3. a. What is the function of the swim bladder?
   to help the perch rise or sink in the water

   b. Why is this an important adaptation for the perch?
   It allows the perch to change position without swimming from one place to another. It also allows the perch to remain at any level in the water without rising or sinking.

Extension

Use forceps to remove one scale from the perch. Place the scale on a microscope slide and cover it with a cover slip. Examine the scale under the low power of a microscope. Count the number of dark rings on the scale. Each ring represents one year's growth. What is the age of the perch?

Name _____ Class _____ Date _____

Chapter 17 — Amphibians and Reptiles

Laboratory 17-1 How do the external and internal structures of a frog adapt it for life both in and out of the water?

Background Information

Amphibians such as frogs are vertebrates that go through metamorphosis. During the early stages of its life cycle, a frog lives in water, has no limbs, and breathes with gills. After metamorphosis, a frog develops powerful limbs and uses lungs to breathe. Although it can live on land, an adult frog feeds and reproduces in water.

Skills: observing, comparing, analyzing

Objectives

In this laboratory, you will
- observe the external and internal structures of a frog.
- identify how the structures are adaptations to life both in and out of the water.

Note to teacher: This laboratory exercise is optional. You may wish to have students identify the structures listed using models, overhead transparencies, blackline masters, computer simulations, and so forth.

Prelab Preparation

Review Section 17-2 Anatomy of a Frog.

Materials

preserved frog   dissecting pan   dissecting needle
dissecting pins   scissors   forceps

Procedure

1. **Caution: Wear gloves during this laboratory.** Rinse a preserved frog under running water to remove as much preservative as possible.

2. Place the frog into a dissecting pan.

3. Observe the appearance of the frog's skin.

4. Locate the frog's mouth, nostrils, eyes, and tympanum.

5. Examine the frog's front and hind legs. Notice the differences between the front and hind legs.

6. Label the following structures on Figure 1: mouth, nostril, eye, tympanum, front leg, hind leg.

7. Turn the frog over so that the ventral surface is facing up. To hold the frog in place, insert a dissecting pin through each of the four legs.

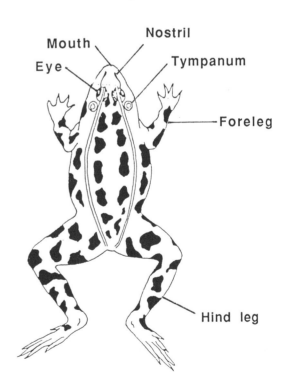

Figure 1: External Structures

Biology   Copyright © by Globe Book Company   83

8. Refer to Figure 2. Pinch a piece of skin in the middle of the frog's ventral surface. Snip off a bit of skin with the scissors. Insert the scissors and cut upward toward the jaw. **Caution: Be careful when using scissors.**

9. Make the other cuts shown in Figure 2 in the correct sequence. This will form two flaps of skin.

10. Cut off the flaps of skin. Underneath the skin is a layer of muscle.

11. Cut through the muscle layer using the same sequence of cuts as you did for the skin.

12. Pull back the muscle layer to expose the internal organs. Use dissecting pins to hold back the two flaps of muscle.

13. If you see a mass of black eggs covering the organs, the frog is a female. Examine the eggs, then cut out the ovaries and remove them from the frog.

14. The uppermost organ is the heart. **Note:** You may have to cut through bone to expose the heart. Carefully cut through the protective sac surrounding the heart. Locate the chambers of the heart.

15. Refer to Figure 3. The largest organ is the liver. Identify the three lobes of the liver.

16. Use forceps to lift the liver. Locate the gallbladder beneath the liver.

17. Locate the stomach. Underneath the stomach is the pancreas.

18. Locate the small intestine. Follow the small intestine to the large intestine. Above the large intestine is the bladder, which empties into the cloaca.

19. Notice that the intestines are held in place by a membrane called mesentery. Cut through the mesentery and straighten the small intestine. Compare the length of the small intestine with the length of the frog.

20. Beneath the digestive organs, locate the reddish-brown kidneys on either side of the backbone.

21. Locate the yellow fat bodies.

22. Above the fat bodies, locate the long, dark red lungs.

23. Below the fat bodies, locate the yellowish testes or white oviducts and ovaries.

24. Describe each of the organs you observed in Data Table 1.

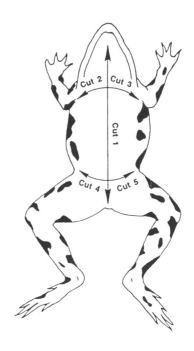

Figure 2: Dissection Technique

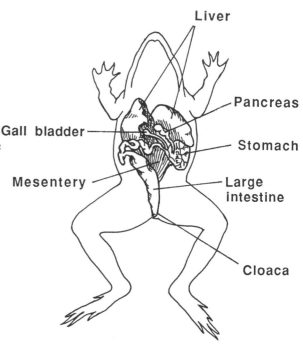

Figure 3: Internal Organs

Name_____ Class_____ Date_____

Chapter 17   Amphibians and Reptiles   Laboratory 17-1

Observations and Data

Data Table 1:   Internal Organs of a Frog

Organs	Description
Heart	three–chambered, inside sac
Liver	large, lobed, reddish–brown
Gallbladder	beneath liver
Stomach	light–colored, saclike
Pancreas	feather–shaped, beneath stomach
Small intestine	long, coiled, held by mesentery
Large intestine	thicker and shorter than small intestine
Bladder	saclike, near large intestine
Cloaca	opening to outside of body
Kidneys	paired, reddish–brown, near backbone
Fat bodies	yellow, fingerlike
Lungs	long, oval, dark red
Testes	yellow
Ovaries	white

Analysis and Conclusions

1. How are the frog's eyes adapted for life in and out of the water?

    The location of the eyes enables the frog to see both in and out of water; the nictitating membranes protect the eyes.

2. How is the frog's skin adapted to life in the water?

    The thin skin can be used for breathing when the frog is in the water.

3. How are the frog's hind legs adapted for life on land and in the water?

    The strong, muscular hind legs are adapted for jumping on land; the webbed feet are adapted for swimming in water.

Biology   Copyright ©   by Globe Book Company   85

Extension

Carefully cut the skin from one of the frog's hind legs. Pull the skin back to expose the muscles. Locate the white cords, or tendons, that attach the muscles to the leg bone. Move the leg back and forth. Observe how the muscles move. Locate the extensor muscles that straighten the leg and the flexor muscles that bend the leg.

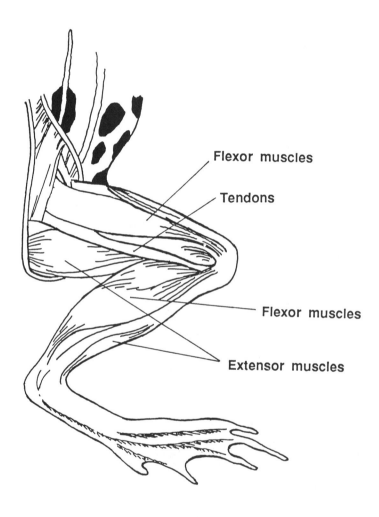

Figure 4   Frog's leg

Name _____ Class _____ Date _____

Chapter 18 — Birds

Laboratory 18-1 How do different bird feathers compare?

Background Information

A body covering of feathers is the characteristic that sets birds apart from other animals. Birds have several different kinds of feathers. Small, fluffy down feathers close to the bird's body insulate the bird and protect it from cold weather. Large contour feathers cover the whole body, give the bird its color, and help it to fly. Filoplumes, or pinfeathers, are long, hairlike feathers. Filoplumes grow in groups near contour feathers and help control the movement of the contour feathers.

Skills: observing, inferring, comparing, contrasting

Objectives

In this laboratory, you will
- observe different kinds of bird feathers.
- compare and contrast the structure and function of different kinds of feathers.

Prelab Preparation

Review Section 18–2 Anatomy of a Bird.

Materials

feathers from different birds
hand lens
probe

Procedure

1. Obtain three different bird feathers.

2. Examine each feather carefully. Compare your feathers with the feathers shown in Figure 1. Identify each feather as a down feather, a contour feather, or a filoplume.

3. Draw each kind of feather in Plates 1, 2, and 3.

4. Label the following structures on your drawings: quill, rachis, vane, barb, barbule.

5. Use a hand lens to observe the vane of each feather.

6. Look closely at the barbs. Use a probe to separate the barbs. Locate the barbules that hook the barbs together.

7. Slowly pull each feather between your thumb and forefinger. Use the hand lens to observe what happens.

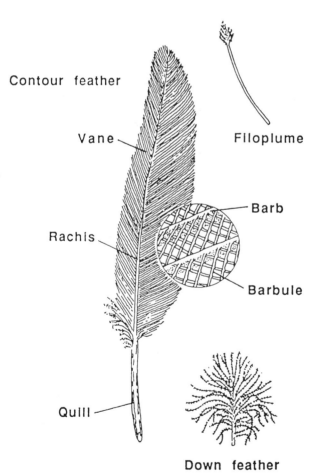

Figure 1: Kinds of Feathers

Biology Copyright © by Globe Book Company 87

Observations and Data

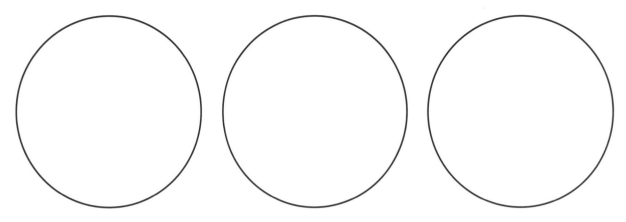

Plate 1: Down Feather   Plate 2: Contour Feather   Plate 3: Filoplume

Analysis and Conclusions

1. How is the structure of a down feather adapted to its function of keeping the bird warm?

   The fluffy shape traps air close to the bird's body and provides a layer of insulation to keep the bird warm.

2. How is the structure of a contour feather adapted for flight?

   The barbs and barbules of a contour feather make the feather airtight and give the bird lift, or upward force, during flight.

3. Why do you think filoplumes are sometimes called pinfeathers?

   because of their shape. Accept all logical responses.

Extension

When a bird preens, it takes oil from a gland at the base of its tail and rubs the oil through its feathers using its beak. Because the oil waterproofs the feathers, preening is especially important for swimming birds such as ducks. What effect do you think detergents in the water might have on the waterproof feathers of a duck? Design an experiment to test your hypothesis.

Name _____ Class _____ Date _____

Chapter 18 — Birds

Laboratory 18-2 How are birds adapted to their environment?

Background Information

The shape of a bird's beak is a good indicator of the kinds of food the bird eats. For example, the beak of a bird that eats seeds is different from the beak of a bird that eats insects. The shape of a bird's feet is adapted to the kind of environment the bird lives in. For example, the webbed feet of a swimming bird are different from the feet of a perching bird. In addition to providing such information about a bird's environment and eating habits, beaks and feet are also useful in identifying birds.

Skills: observing, comparing, contrasting

Objectives

In this laboratory, you will
- observe birds in their natural environment.
- compare the birds you observe with pictures of birds in a field guide to identify the birds.

Prelab Preparation

1. Practice using binoculars by focussing on different objects; become familiar with the adjustment knobs.

2. Look through a field guide to birds; become familiar with the appearance of some common species, such as sparrows, woodpeckers, pigeons, and crows.

Materials

binoculars
notebook
pencil
field guide to birds

Procedure

1. You will work in groups of three: one student will be the observer, another will be the recorder, and the third will be the field guide reader. Students should take turns performing these tasks.

2. Your teacher will take you on a field trip to the school yard, a nearby park or garden, or other natural area.

3. Observers will look for birds on the ground, in the water, or in the trees. They should use the binoculars to observe as many details of the beaks and feet as possible.

4. Recorders should record observations in Data Table 1.

5. Field guide readers should try to identify the birds observed using the pictures and descriptions in the field guide.

Biology Copyright © by Globe Book Company

Observations and Data

Data Table 1: Characteristics of Beaks and Feet

Birds	Beaks	Feet
1	Answers will vary.	
2		
3		
4		
5		

Analysis and Conclusions

1. Using the field guide, identify each of the birds you listed in Data Table 1. Use your observations to support your answers.

   Answers will vary.

2. Why is a long back toe on the foot of a perching bird an important adaptation?

   It helps the bird stay on its perch by wrapping around the perch.

3. How is a short, stout beak useful to a seed eater?

   by helping the bird crack open seeds

4. Why is a strong pointed beak an important adaptation for woodpeckers?

   woodpecker searches for insects

Extension

If possible, visit a zoo and observe several birds with unusual beaks, such as flamingoes, spoonbills, and hummingbirds. Describe how each bird's beak is adapted to the kind of food it eats. Record your observations in a data table.

Name_____ Class_____ Date_____

Chapter 19   Mammals

Laboratory 19-1 What kinds of tracks do animals leave?

Background Information

Animals leave tracks when they walk through mud or snow. Each animal leaves a distinctive track. Naturalists, or people who have experience with wildlife, can identify the kind of animal that left a particular track.

Skills: observing, comparing and contrasting

Objective

In this laboratory, you will
- observe and identify different kinds of animal tracks.

Prelab Preparation

Study the diagram of animal tracks shown in Figure 1. Note the characteristics that make one set of tracks appear different from the others.

Materials

shallow pan	paper towels
soil	water
scissors	pencil

Note to teacher: Divide the class into groups of two students and assign each group a number.

Procedure

1. You will work in groups of two students. Mix some soil with a small amount of water in a shallow pan. Be sure not to use too much water. Smooth over the top of the muddy soil.

2. Use scissors to cut out one of the animal tracks in Figure 1. **Caution: Be careful when using scissors.**

3. Trace the footprint cut outs onto a piece of styrofoam or foam core.

4. Using the lines you traced in step 3, carefully cut out the track from the styrofoam or foam core.

5. Lightly press the cut-out track into the muddy soil in your pan. If you have space in the pan, make two impressions of the track. Trace the outline of the track in the muddy soil with a pencil before removing the track from the pan.

6. Remove the track from the pan. Carefully press the track into the muddy soil to make a clear impression about 1 cm deep. Use your fingers or a pencil.

7. Label your pan with the number of your group.

8. Observe the tracks made by the other groups in your class. In Data Table 1, identify which animal you think matches each track.

Biology   Copyright ©   by Globe Book Company

Observations and Data

Data Table 1: Animal Tracks

Group Number	Animal
1	Answers will vary.
2	
3	
4	
5	

Analysis and Conclusions

1. What are some differences between the tracks?

   Answers will vary. Possible responses might include placement of toes, split or unsplit hoof, etc.

   Accept all logical responses.

2. How could you preserve your tracks for later study?

   by making plaster of Paris casts

3. How do you think a naturalist could estimate by observing tracks how fast an animal was moving when it left the tracks?

   Answers will vary. The distance between the tracks increases as the animal moves faster; the relative position of the front and back tracks changes. Accept all logical responses.

Extension

Use reference materials to research other kinds of animal tracks. Create a bulletin board display of at least four different kinds of animal tracks. Identify each animal track and include a picture of the animal that made each track.

Name _____ Class _____ Date _____

Chapter 19   Mammals   Laboratory 19-1

Black Bear

Beaver

Moose

Raccoon

Figure 1:   Animal Tracks

Biology Copyright © by Globe Book Company   93

Name _____ Class _____ Date _____

Chapter 19 | Mammals

Laboratory 19-2 How are placental mammals classified?

Background Information

Biologists classify placental mammals according to the physical characteristics that help the animal adapt to its environment. Eight orders of placental mammals are rodentlike mammals, insect-eating mammals, gnawing mammals, aquatic mammals, trunk-nosed mammals, carnivorous mammals, hoofed mammals, and primates.

Skills: observing, classifying, comparing and contrasting

Objective

In this laboratory, you will
- use a dichotomous key to classify eight kinds of placental mammals.

Prelab Preparation

1. Review Section 19–3 Placental Mammals.

2. Review how to use a dichotomous key.

Materials

paper pencil

Procedure

1. Study the animals shown in Figure 1.

2. Use the dichotomous key shown in Figure 2 to identify the order of placental mammals to which animal A belongs. Record your identification in Data Table 1.

3. Repeat steps 1 and 2 for the remaining animals in Figure 1.

Observations and Data

Data Table 1: Placental Mammals

Animal	Identification
A (sea cow)	aquatic
B (monkey)	primate
C (rabbit)	rodentlike
D (elephant)	trunk–nosed
E (porcupine)	gnawing
F (lion)	carnivorous
G (shrew)	insect-eating
H (camel)	hoofed

Biology   Copyright ©   by Globe Book Company

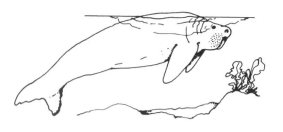
A (sea cow)

B (monkey)

C (rabbit)

D (elephant)

E (porcupine)

F (lion)

G (shrew)

H (camel)

Figure 1: Placental Mammals

Name _____ Class _____ Date _____

Chapter 19 Mammals

Laboratory 19-2

Figure 2:   Dichotomous Key for Placental Mammals

Questions	Yes	No
1. Does the animal breathe with lungs?	Go to 2.	Go to 10.
2. Does the animal live in water?	This is an aquatic mammal.	Go to 3.
3. Does the animal have opposable thumbs?	This is a primate.	Go to 4.
4. Does the animal eat plants?	Go to 5.	Go to 6.
5. Does the animal have hooves?	This is a hoofed mammal.	Go to 7.
6. Does the animal have canine teeth?	This is a carnivorous mammal.	Go to 9.
7. Does the animal have four upper incisors?	This is rodentlike mammal.	Go to 8.
8. Does the animal have a trunk?	This is a trunk–nosed mammal.	Go to 10.
9. Does the animal eat insects?	This is an insect–eating mammal.	Go to 10.
10. Does the animal have two upper incisors?	This is a gnawing mammal.	If you have not yet identified the animal, return to 1.

Biology   Copyright ©   by Globe Book Company

Analysis and Conclusions

1. How is a dichotomous key useful in identifying different kinds of placental mammals?

   By answering a series of questions about an animal's physical characteristics, it is possible to identify to which order of placental mammals the animal belongs.

2. Which order of placental mammals has an opposable thumb?

   primates

3. Based on the dichotomous key in Figure 2, which characteristic is shared by all orders of placental mammals?

   They all breathe with lungs.

Extension

Develop a dichotomous key to identify members of the three major groups of mammals, based on the way in which the young develop.

Name _____ Class _____ Date _____

Chapter 20   Support and Movement

Laboratory 20-1 What kinds of movements are possible for human body joints?

Background Information

Movement is possible because of the way in which your bones and muscles are arranged. Because muscles can only pull, two muscles working in opposing pairs are necessary to bend and straighten a bone at a joint. Bending a joint is called flexion; straightening a joint is called extension. Lifting an arm or leg away from the body is called abduction; bringing it back toward the body is called adduction.

Skills: modeling, observing, comparing, analyzing

Objective

In this laboratory, you will
- observe the location of muscles that make various kinds of movement possible.

Prelab Preparation

1. Review Section 20-3 The Muscular System.

2. Review the meaning of the prefixes "ex-," "ab-," and "ad."

   "ex-" means out; "ab-" means from or away; "ad-" means toward

Materials

paper
pencil

Procedure

1. Carefully bend your head backward at the neck. Refer to Figure 1.

2. Gently feel the muscles at the back and sides of your neck that make this movement possible. See Data Table 1.

3. Carefully bend your head forward, as shown in Figure 2.

4. Gently feel the muscles at the back and sides of your neck that make this movement possible.

5. Refer to Figure 3. Carefully imitate the movement shown in diagram A. Locate the muscles that make this movement possible. Record the location of the muscles and the kind of movement (flexion, extension, abduction, or adduction) in Data Table 2.

6. Carefully imitate each of the other movements shown in Figure 3. Locate the muscles responsible for each movement. Record the location of the muscles and the kind of movement in Data Table 2.

Figure 1

Figure 2

Biology   Copyright © by Globe Book Company   99

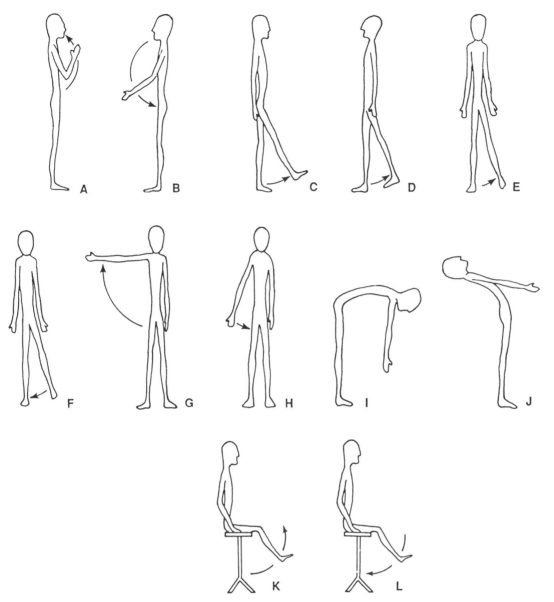

Figure 3: Body Movements

Observations and Data

Data Table 1: Head Movements

Movement	Location of Muscles	Kind of Movement
Backward	back of neck	extension
Forward	neck (sides of chin)	flexion

Name _____ Class _____ Date _____

Chapter 20 Support and Movement   Laboratory 20-1

Data Table 2: Body Movements

Movement	Location of Muscles	Kind of Movement
A	front upper arm	flexion
B	back upper arm	extension
C	front thigh	flexion
D	buttocks	extension
E	outer thigh	abduction
F	inner thigh	adduction
G	top shoulder	abduction
H	back and chest	adduction
I	abdomen	flexion
J	back	extension
K	front upper leg	extension
L	back upper leg	flexion

1. Refer to Data Table 1. What kind of movement is involved in bending your head backward?
   extension

2. What kind of movement is involved in bending your head forward?
   flexion

3. Using Data Table 2, list two examples of flexion and extension.
   Answers will vary.

4. List two examples of abduction and adduction.
   Answers will vary.

Biology   Copyright © by Globe Book Company   101

Analysis and Conclusions

1. What kind of movement is involved when you put up your hand to "stop traffic"? How do you know?

   abduction, because you move your arm away from your body

2. What kinds of movement are involved in walking?

   flexion and extension

3. The four kinds of movement you observed in this laboratory can be combined in a movement called circumduction. Pitching a baseball involves circumduction. What other movement can you name that involves circumduction?

   swinging your leg in a circle

Extension

Use reference materials to find out what a chiropractor does to help people with joint or muscle problems. Present your results in a written report. Do you think the work of chiropractors is helpful or harmful? Explain your answer.

Name _____ Class _____ Date _____

Chapter 21　　　　　　　　　　　　　　　　　　　　　　　　　Digestion

Laboratory 21-1 How does saliva help in the chemical digestion of food?

Background Information

Saliva is produced by glands under the tongue, behind the jaw, and in front of the ear. These glands are connected to the mouth by tubes, or ducts. Saliva is made up of water, mucus, and an enzyme called ptyalin. Ptyalin (amylase) helps begin the chemical digestion of starch into sugar.

Skills: observing, analyzing

Objective

In this laboratory, you will
- use indicators to observe the effect of ptyalin on starch.

Prelab Preparation

1. Review Section 21-1 The Digestion Process and Section 21-2 The Digestive System.

2. Review the proper technique for lighting and using a Bunsen burner.

3. a. What is an indicator?

   a substance that changes color in the presence of a certain chemical

   b. What is an enzyme?

   a protein that is involved in chemical reactions in the body

Materials

starch solution	Lugol's solution	Benedict's solution
3 test tubes	test tube rack	ptyalin (amylase) solution
test tube holder	dropper	Bunsen burner
marking pencil		

Procedure

1. Place three clean, dry test tubes in a test tube rack. Number the tubes 1 to 3.

2. Add 3 mL of starch solution to each of the test tubes.

3. Add three drops of Lugol's solution to each test tube. Swirl gently to mix the liquids. A color change to blue-black indicates the presence of starch. Record your observations in Data Table 1.

4. Add 2 mL of ptyalin solution to test tube 3. Swirl gently to mix the liquids.

5. Add 3 mL of Benedict's solution to test tubes 2 and 3. Swirl gently to mix the liquids. **Caution: Benedict's solution can irritate the skin. Wash with cold water immediately if any spills on your skin.**

6. Using a test tube holder, gently heat the solution in test tube 2 over a Bunsen burner. **Caution: Be careful when using a Bunsen burner.** A color change to green, yellow, orange, or red indicates the presence of sugar. Record your observations in Data Table 1.

7. Repeat step 6 with test tube 3.

Biology　　　　　　　　　　Copyright ©　　by Globe Book Company

Observations and Data

Data Table 1: Tests for Starch and Sugar

Test Tube	Indicator	Color Change
1	Lugol's	blue-black
2	Lugol's	blue-black
	Benedict's	none
3	Lugol's	blue-black
	Benedict's	red

1. What happened to the starch solution in the three test tubes when you added Lugol's solution?
   The starch solution turned blue-black.

2. What happened to the solution in test tube 2 after you added Benedict's solution and heated the test tube?
   The color did not change.

3. What happened to the solution in test tube 3 after you added Benedict's solution and heated the test tube?
   The color changed from blue to yellow to red.

Analysis and Conclusions

1. What did adding Lugol's solution to test tubes 1–3 demonstrate?
   the presence of starch

2. What did the color change in test tube 3 indicate?
   the presence of sugar

3. Why was sugar present in test tube 3, but not in test tube 2?
   Ptyalin was added to test tube 3; the ptyalin changed the starch in test tube 3 into sugar.

Extension

Does heat affect the ability of ptyalin to change starch into sugar? Design an experiment to test the effect of heat on saliva. Record your results in a data table.

Name _____ Class _____ Date _____

Chapter 22 | Circulation

Laboratory 22-1 What kinds of cells are found in human blood?

Background Information

Blood is a complicated liquid tissue. The liquid part, or plasma, makes up 55 percent of blood. The remaining 45 percent of blood is made up of three different kinds of cells: red blood cells, white blood cells, and platelets. Red blood cells are the most numerous. There are usually five kinds of white cells. Platelets are tiny, colorless fragments of large cells made by the bone marrow.

Skills: observing, comparing

Objective

In this laboratory, you will
- observe and compare the size, appearance, and number of blood cells.

Prelab Preparation

1. Review Section 22-3 Blood.

2. Review the proper procedure for using a compound microscope.

Materials

compound microscope
prepared slide of human blood

Procedure

1. Place a prepared slide of human blood on the stage of a compound microscope.

2. Use the low-power objective to focus on the slide. Then carefully switch to high power.

3. The blood on the prepared slide has been stained. Red blood cells will appear pink; the nuclei of white blood cells will appear blue-purple; the platelets will appear as small blue specks.

4. Search the slide slowly and carefully to locate red blood cells. Sketch several red blood cells in Plate 1.

5. Different kinds of white blood cells can be identified by the shape of their nuclei and the way in which the cytoplasm absorbed the stain. Search the slide and locate as many different kinds of white blood cells as you can. Sketch these white blood cells in Plate 2.

6. Search the slide to locate platelets. Sketch the platelets in Plate 3.

7. Complete Data Table 1.

Observations and Data

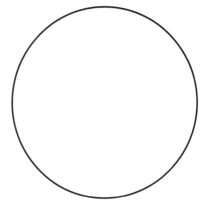
Plate 1: Red Blood Cells

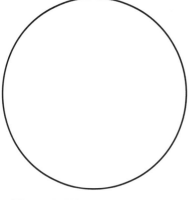
Plate 2: White Blood Cells

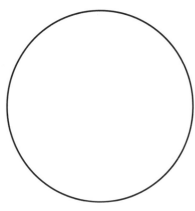
Plate 3: Platelets

Data Table 1: Blood Cells

Appearance	Red Blood Cells	White Blood Cells	Platelets
Shape	disklike	roundish	irregular
Color	pink	white cytoplasm, blue nuclei	blue
Size	small	large	tiny

Analysis and Conclusions

1. White blood cells are often described as "amoeboid." What does this term mean? Why is it applied to white blood cells?

    "Amoeboid" refers to amoebas; it is applied to white blood cells because they resemble amoebas and move

    by amoeboid motion (pseudopods).

2. Why do you think amoeboid motion is an advantage to white blood cells?

    Accept all logical answers. They can squeeze through small spaces as they move throughout the body to

    fight infections.

3. What do you think a doctor might conclude if a sample of a patient's blood had a higher than normal number of white blood cells?

    that the patient has an infection

Extension

Use reference materials to find out what a blood count is and how it is performed. What can a doctor learn from a blood count? What is a differential blood count? What diseases can be diagnosed by studying a person's blood? Present your findings in a written report.

Name_____ Class_____ Date_____

Chapter 22 | Circulation

Laboratory 22-2 Which blood types can be mixed safely?

Background Information

The differences among the four major blood types are caused by antigens on the surface of red blood cells. Type A blood has A antigens and type B blood has B antigens. Type AB blood has both A and B antigens, while type O blood has neither antigen. Each blood type also has antibodies. Antibodies are chemicals that attack foreign substances in the blood. For example, type A antibody will attack type B antigen. If a person with type A blood receives a transfusion of type B blood, the red blood cells in the two blood types will clump together. These clumps can block blood vessels and may lead to death.

Skills: modeling, predicting, observing

Objectives

In this laboratory, you will
- predict which blood types can be mixed together safely.
- use a model to observe which blood types can be mixed together safely.

Prelab Preparation

Review Section 22-3 Blood.

Materials

8 paper cups	red food coloring	blue food coloring
water	dropper	marking pencil

Procedure

1. Label four paper cups A, B, AB, and O.

2. Fill each of the four cups half full of water.

3. Add two drops of red food coloring to cup A; two drops of blue food coloring to cup B; two drops of red and two drops of blue food coloring to cup AB. Do not add any food coloring to cup O.

4. The cups labeled A, B, AB, and O represent four "patients" with four different blood types who are in need of a blood transfusion. Which of the four blood types can be safely mixed with each patient's blood type? Record your predictions in Data Table 1. Use the words "safe" or "unsafe."

5. Label four other paper cups A', B', AB', and O'. These cups will represent the blood types available for transfusion.

6. Repeat steps 2 and 3 with cups A', B', AB', and O'.

7. Using a clean dropper, transfer two drops of "blood" from cup A' to "patient" cup A. If the patient's blood does not change color, the transfusion is safe. If the patient's blood does change color, the transfusion is unsafe. Record your results in Data Table 1.

8. Repeat step 7 until each of the blood types listed in Data Table 1 have been mixed. Be sure to clean the dropper after each use. Record your results in Data Table 1.

Biology Copyright © by Globe Book Company 107

Observations and Data

Data Table 1: Results of Transfusions

Blood Type (Patient)	Blood type (Transfusion)							
	A		B		AB		O	
	predicted	actual	predicted	actual	predicted	actual	predicted	actual
A		safe		unsafe		unsafe		safe
B		unsafe		safe		unsafe		safe
AB		safe		safe		safe		safe
O		unsafe		unsafe		unsafe		safe

1. Which blood type(s) can a person with type A blood safely receive?
   A, O

2. Which blood type(s) can a person with type B blood safely receive?
   B, O

3. Which blood type(s) can a person with type AB blood safely receive?
   A, B, AB, O

4. Which blood type(s) can a person with type O blood safely receive?
   O

5. Were your predictions correct?
   Answers will vary.

Analysis and Conclusions

1. Why do you think people with type AB blood are called "universal recipients"? Use your observations to support your answer.
   They can safely receive all four blood types; their red blood cells have neither A nor B antibodies.

2. Why do you think people with type O blood are called "universal donors"? Use your observations to support your answer.
   They can safely donate blood to all four blood types; their red blood cells have neither A nor B antigens.

Extension

A blood transfusion can be either whole blood or plasma. Use reference materials to find out the difference between whole blood and plasma. What are the advantages and disadvantages of each?

Name _____ Class _____ Date _____

Chapter 23 — Respiration and Excretion

Laboratory 23-1 How much air do you normally exhale?

Background Information

When you breathe normally, you inhale about 500 mL of air into your lungs. This amount of air is called tidal volume. You also can inhale an additional volume of air called the reserve volume. After exhaling normally, you can forcibly exhale this same additional volume of air. Even after the most forceful exhalation, however, about 1000 mL of air remains in your lungs. This amount of air is called the residual volume.

Skills: modeling, observing, recording data

Objective

In this laboratory, you will
- measure the amount of air you normally exhale.

Prelab Preparation

1. Review Section 23-2 Breathing.

2. Review how to calculate an average.

Note to teacher: You may wish to insert the glass tubing into the rubber stoppers before students begin the laboratory. If students insert the tubing themselves, be sure that the tubing is well lubricated, that the ends are fire polished, and that students wear gloves.

Materials

2-L plastic bottle
food coloring
2 pieces of glass tubing, 10 cm and 20 cm long
3 30-cm pieces of rubber tubing
water
2-hole rubber stopper
250-mL beaker
100-mL graduated cylinder

Procedure

1. Fill a 2-L plastic bottle about three-fourths full of water. Add a few drops of food coloring to the water.

2. Carefully insert two pieces of glass tubing into a 2-hole rubber stopper. **Caution: Be careful when inserting the glass tubing; wear gloves.**

3. Place the rubber stopper and glass tubing into the plastic bottle. Be sure that the end of the longer glass tubing is below the surface of the water. The end of the shorter piece of glass tubing must be above the surface of the water. See Figure 1.

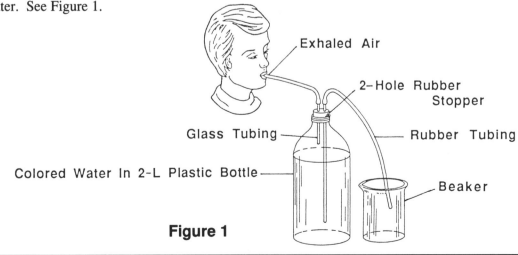

Figure 1

Biology — Copyright © by Globe Book Company — 109

4. Carefully attach one end of a piece of rubber tubing to the longer piece of glass tubing. **Caution: Be careful when attaching the rubber tubing; use a slow, twisting motion.**

5. Place the free end of the rubber tubing into a 250-mL beaker.

6. Attach a second piece of rubber tubing to the shorter piece of glass tubing.

7. Inhale normally. Then exhale normally through the free end of the rubber tubing attached to the shorter piece of glass tubing. Refer to Figure 1. The air you exhale will displace water from the bottle, through the tubing, and into the beaker.

8. The volume of water displaced is equal to the volume of air you exhaled. Measure the volume of water by pouring the water from the beaker into a graduated cylinder. Record this volume as Trial 1 in Data Table 1.

9. Pour the water from the graduated cylinder back into the 2-L bottle.

10. Repeat steps 7 to 9 two more times. Record your results as Trials 2 and 3 in Data Table 1.

11. Calculate the average volume of water displaced by adding the results of the three trials and dividing by three. Record the average volume in Data Table 1.

12. Replace the piece of rubber tubing you exhaled into with a new piece of rubber tubing.

13. Have your partner repeat steps 7 to 11. Complete Data Table 1 with your partner's results.

Observations and Data

Data Table 1: Volume of Water Displaced

Volume	Partner 1	Partner 2
Trial 1	**Answers will vary.**	
Trial 2		
Trial 3		
Average		

1. Who exhaled the greater average volume of air—you or your partner?
   Answers will vary.

2. How does the volume of air you exhaled compare with the volume of air your partner exhaled in each trial?
   Answers will vary.

Name _____ Class _____ Date _____

Chapter 23 Respiration and Excretion   Laboratory 23-1

Analysis and Conclusions

1. How can you explain any difference in the average volume of air exhaled by you and by your partner?

   Answers will vary. Body size and sex will affect the volume exhaled.

2. Do you think the average volume of air you exhale will increase or decrease after exercise? Explain your answer.

   Accept all logical answers. The average volume exhaled may decrease, but the

   rate of respiration will increase.

3. Count the number of breaths you normally take in one minute. Then calculate the average volume of air you exhale in one minute. (Hint: Multiply the number of breaths you take in one minute by the average volume of air you exhale in one breath.)

   Answers will vary.

4. How much air do you exhale in one class period? Show your calculations.

   Answers will vary.

Biology   Copyright ©   by Globe Book Company   111

Extension

Design an experiment to answer question 2 in the Analysis and Conclusions section. Organize your results in a data table that compares the average volume of air exhaled before exercise with the average volume of air exhaled after exercise. Use the space provided to describe your experiment.

Name _____ Class _____ Date _____

Chapter 23   Respiration and Excretion

Laboratory 23-2 What effect does exercise have on the amount of carbon dioxide you exhale?

Background Information

Carbon dioxide (CO_2) is a waste product of respiration. As you exercise, your respiration rate increases to supply the extra energy your body needs to continue exercising. As a result, the amount of carbon dioxide in your blood also increases. The more strenuous the exercise, the more carbon dioxide is produced.

Skills: observing, analyzing

Objective

In this laboratory, you will
- use an indicator to find out how exercise affects the amount of carbon dioxide you exhale.

Prelab Preparation

1. Review Section 23-2 Breathing.

2. Review the function of an indicator.

Materials

bromthymol blue solution	graduated cylinder	dropper
dilute ammonia solution	250-mL beaker	2 drinking straws
stirring rod	watch or clock with second hand	

Procedure

1. **Caution: Bromthymol blue and ammonia can stain clothing and irritate skin. Be sure to wear your lab apron and goggles during this laboratory.** Pour 100 mL of bromthymol blue solution into a 250 mL beaker.

2. Exhale slowly through a drinking straw into the bromthymol blue for 1 minute, stopping to inhale as necessary. **Caution: Be sure to take the straw out of your mouth when you inhale.**

3. The bromthymol blue should turn yellow as you exhale into it. If the solution does not turn yellow after 1 minute, continue exhaling into the solution until it changes color.

4. Add one drop of ammonia to the yellow solution and stir once with a stirring rod.

5. Continue to add ammonia one drop at a time, stirring once after each drop, until the solution turns blue.

6. Record the number of drops of ammonia you added under Trial 1, Before Exercise, in Data Table 1.

7. Repeat steps 2 to 5 two more times. Record your results as Trial 2 and Trial 3, Before Exercise, in Data Table 1.

8. Calculate the average number of drops of ammonia needed to change the color of the bromthymol blue solution from yellow to blue. Record the average in Data Table 1.

9. Carefully exercise by running in place for 1 minute.

10. Repeat steps 2 to 8. Record your results in the After Exercise column of Data Table 1.

Biology   Copyright © by Globe Book Company   113

Observations and Data

Data Table 1: Results of Exercise on CO_2

Drops of Ammonia	Before Exercise	After Exercise
Trial 1	Answers will vary.	
Trial 2		
Trial 3		
Average		

1. What was the average number of drops of ammonia needed to change the color of the solution from yellow to blue before exercise?

   Answers will vary.

2. What was the average number of drops needed after exercise?

   Answers will vary. The number of drops should increase after exercise.

3. Compare your results with two other members of your class. Were the averages the same or different?

   Answers will vary.

Analysis and Conclusions

1. How is the average number of drops of ammonia related to the amount of carbon dioxide exhaled?

   The more carbon dioxide exhaled, the greater the average number of drops of ammonia needed to change the color of the solution.

2. How does exercise affect the amount of carbon dioxide you exhale? How do you know? Use your data to support your answer.

   The amount of carbon dioxide exhaled increases after exercise. The number of drops of ammonia needed to change the color of the solution increased after exercise.

3. What was the control in this laboratory?

   the "Before Exercise" results

4. List two variables that might have affected the results of this laboratory.

   Accept all logical answers. Body size, sex, and physical condition are some possible answers.

Extension

Design an experiment to find out if seeds give off carbon dioxide during respiration. Record your results in a data table.

Name _____ Class _____ Date _____

Chapter 24 — Regulation

Laboratory 24-1 Which side of your brain is dominant?

Background Information

The functions of the cerebrum, the upper portion of the brain, are divided between its two halves. The right half of the cerebrum controls the movement of muscles on the left side of the body. The left half controls movement on the right side of the body. Language comprehension is controlled by the left half of the cerebrum; feelings and emotions are controlled by the right half.

Skills: observing, analyzing, inferring

Objective

In this laboratory, you will
- perform various tasks to find out which side of your brain is dominant.

Prelab Preparation

1. Review Section 24-1 The Nervous System.

2. Review the definition of the word "dominant."
    commanding or controlling

3. Read through the steps listed in the Procedure.

4. Design a data table to record your observations. The title should be "Data Table 1: Finding the Dominant Side of the Body." Across the top of the table, write the following headings: Task, Right, Left. Under the heading "Task," list each of the tasks described in the Procedure.

Materials

paper
pencil

Procedure

1. Work with a partner. Have your partner perform each of the following tasks. Observe which hand your partner uses, and place a check mark in the "Right" or "Left" column of Data Table 1.
    a. pick up a pencil from the desk
    b. write your name
    c. pick up a pencil from the floor at the right side of your chair
    d. pick up a pencil from the floor at the left side of your chair
    e. turn the pages of a book

2. Have your partner fold his/her hands. Observe which thumb is on top and place a check mark in the "Right" or "Left" column of Data Table 1.

3. Observe which foot your partner uses for each of the following tasks. Place a check mark in the proper column of Data Table 1.
    a. stand on one foot
    b. take one step forward
    c. walk up one step of a staircase
    d. walk down one step of a staircase

Biology

4. Roll a sheet of paper into a tube. Have your partner look through the tube, with both eyes open, at an object on the opposite side of the classroom.

5. Have your partner close first the right eye and then the left. Ask your partner with which eye he/she could see the object through the tube. Record your partner's response in Data Table 1.

6. Add up the number of check marks in each column of Data Table 1 and record the totals for each column.

7. Change places with your partner and repeat the procedure.

Observations and Data

1. Which column in Data Table 1, Right or Left, had more check marks?
   Answers will vary.

2. Which side of your body is dominant?
   Answers will vary.

3. Compare your data with the data collected by other members of your class. a. How many students are right-handed and have a right-side dominance?
   Answers will vary.

   b. How many are left-handed and have a left-side dominance?
   Answers will vary.

Analysis and Conclusions

1. Which side of your brain is dominant? How do you know?
   Answers will vary. Students with right-side dominance are left-brained; those with left-side dominance are right-brained.

2. What percent of your class have the same dominant side of the brain?
   Answers will vary.

3. Is it possible for a person who is right-handed to have right-brain dominance?
   Accept all logical answers. Some people who have right-brain dominance and would naturally be left-handed might have been trained to be right-handed.

Extension

Use reference materials to find out what causes a person to be left-handed or right-handed. What differences are there between "lefties" and "righties"? What are the advantages and disadvantages of being a "lefty" or a "righty"? Present your findings in a written report.

Name_____ Class_____ Date_____

Chapter 24 — Regulation

Laboratory 24-2 How can you test your senses of touch and hearing?

Background Information

You receive information from the environment through special sensory receptors. These receptors transmit messages to your brain through sensory neurons. The receptors for touch, pressure, and pain are scattered throughout the skin; the receptors for sound are located in the inner ear.

Skills: observing, sequencing, analyzing

Objectives

In this laboratory, you will
- locate touch receptors on your skin.
- test the limits of your sense of hearing.

Prelab Preparation

Review Section 24-3 The Five Senses.

Materials

index card	meterstick	9 dissecting pins
tuning fork	scissors	rubber hammer
metric ruler	cotton balls	

Procedure

1. On an index card, draw five rectangles, each 4 cm x 6 cm. Number the rectangles 1 to 5.

2. Draw two diagonal lines on each rectangle to locate the center of the rectangle. See Figure 1.

3. On rectangle 1, place a dot in the center, as shown in Figure 1. On rectangle 2, place one dot in the center and one dot 5 mm from the center. On rectangle 3, place one dot in the center and one dot 10 mm from the center. On rectangle 4, place one dot in the center and one dot 15 mm from the center. On rectangle 5, place one dot 15 mm to one side of the center and a second dot 15 mm to the opposite side of the center. The two dots should be 30 mm apart, as shown in Figure 2.

4. Carefully cut out each of the rectangles. **Caution: Be careful when using scissors.**

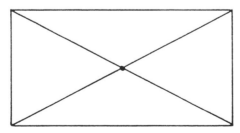

Figure 1

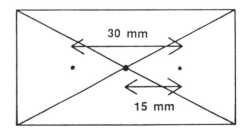

Figure 2

Biology — Copyright © by Globe Book Company — 117

5. Carefully insert a dissecting pin through each of the dots on rectangles 1–5.

6. Work with a partner. Have your partner close his/her eyes.

7. Gently, without using any pressure, touch the tip of the pin in rectangle 1 to your partner's fingertip. If your partner feels the touch of the pin, place a + under the heading One Pin in Data Table 1. If your partner does not feel the pin, place a – in Data Table 1.

8. Repeat step 7 for each of the body parts listed in Data Table 1.

9. Use rectangle 2 to touch two pins to your partner's fingertip. Repeat for each body part and record your results in Data Table 1. If your partner feels two pins, place a + in the correct column; if he/she feels only one pin, place a – in the correct column.

10. Repeat step 9 using rectangles 3, 4, and 5.

11. Change places with your partner and repeat steps 7–10.

12. To test your partner's sense of hearing, strike a tuning fork and hold it near your partner's left ear. Slowly move away until your partner can no longer hear the tuning fork.

13. Measure the distance from your partner's ear to the point where he/she could no longer hear the tuning fork. Record this distance in Data Table 2.

14. Repeat steps 12 and 13 for your partner's right ear.

15. Have your partner alternately block his/her left and right ear with cotton. Repeat steps 12–14.

16. Have your partner close his/her eyes. Strike the tuning fork 1 m in front of your partner. Ask him/her from which direction the sound came. Record a + for a correct response, and a – for an incorrect response, in Data Table 3.

17. Repeat step 16 for each of the directions listed in Data Table 3.

18. Have your partner alternately block his/her left and right ear with cotton. Repeat steps 16 and 17.

19. Change places with your partner and repeat steps 12–18.

Observations and Data

Data Table 1: Sense of Touch

Body Part	One Pin	Two Pins (distance apart)			
		(5 mm)	(10 mm)	(15 mm)	(30 mm)
Fingertip			Answers will		
Palm			vary.		
Back of hand					
Back of neck					
Inside of forearm					

Name_____ Class_____ Date_____

Chapter 24 Regulation

Laboratory 24-2

Data Table 2: Sense of Hearing

Trial	Distance (m)	
	Left ear	Right ear
Both ears open	Answers will vary.	
Left ear blocked		
Right ear blocked		

Data Table 3: Direction of Sound

Direction	Both ears open	Left ear blocked	Right ear blocked
Front	Answers will vary.		
Back			
Left			
Right			
Above			

1. Which body part was most sensitive to touch?
   fingertips

2. How does your sense of hearing compare with your partner's?
   Answers will vary.

3. Exchange data with other members of your class. What was the greatest distance at which the tuning fork could be heard? Was this with both ears open or with one ear blocked?
   Answers will vary. Students will probably report for both ears open.

Biology

Analysis and Conclusions

1. Why are some areas of the skin more sensitive to touch than others?
   They have more receptors.

2. Why are the fingertips more sensitive than other body parts?
   because of the need to touch and manipulate objects with the fingers

3. How does distance affect your sense of hearing? Use your data to support your answer.
   Sense of hearing decreases as distance increases.

Extension

The taste buds on your tongue are sensitive to only four tastes: sweet, sour, salty, and bitter. Design an experiment to locate the various taste receptors on your tongue. Record your observations in a data table. Use the space provided to describe your experiment.

Name _____ Class _____ Date _____

Chapter 24 | Regulation

Laboratory 24-3 How can your senses be fooled?

Background Information

What you think you are seeing is sometimes very different from what you are actually seeing. You see with your eyes. Then your brain interprets what you see. But the brain can be fooled into misinterpreting what you see. Something that is not what it at first appears to be is called an optical illusion.

Skills: observing, measuring, predicting

Objective

In this laboratory, you will
- compare what your eyes are seeing with what your brain tells you you are seeing.

Prelab Preparation

1. Review Section 24-3 The Five Senses

2. What is an optical illusion?

   a misleading image presented to the sight

Materials

metric ruler   paper   pencil

Procedure

1. Look at Figure 1. Which line, A or B, appears longer? Record your prediction in Data Table 1.

2. Measure the length of each line in Figure 1. Record which line is actually longer in Data Table 1.

3. Look at Figure 2. Which appears greater, the distance between A and B or the distance between B and C? Record your prediction in Data Table 1.

4. Measure the distance from A to B and from B to C. Record which distance is actually greater in Data Table 1.

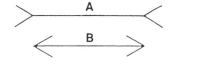

Figure 1   Figure 2

Biology   Copyright © by Globe Book Company   121

5. Look at Figure 3. Does the diagonal line appear straight or bent? Record your prediction in Data Table 1.

6. Use a metric ruler to determine if the line in Figure 3 is actually straight or bent. Record the result in Data Table 1.

7. Look at Figure 4. Which of the three cylinders — A, B, or C — appears tallest? Record your prediction in Data Table 1.

8. Measure the height of each cylinder in Figure 4. Record which cylinder is actually the tallest in Data Table 1.

9. Look at Figure 5. Which surface of the cube appears to be shaded? Record your observation under "First Look" in Data Table 2.

10. Look at Figure 5 again. Does the same surface still appear shaded? Record your observation under "Second Look" in Data Table 2.

11. Look at Figure 6. Record what you see in Data Table 2.

12. Look at Figure 6 again. Do you see something different? Record your observation in Data Table 2.

13. Repeat steps 11 and 12 for Figures 7 and 8.

Observations and Data

Data Table 1: Length/Distance

Figure	Prediction	Actual
1	B is longer	A and B are equal
2	A to B is greater than B to C	both are equal
3	line is bent	line is straight
4	C is tallest	A, B, and C are equal

Data Table 2: Appearances

Figure	First Look	Second Look
5		
6	Answers will vary.	
7		
8		

Note to teacher: In Figure 5, students should see either the top or back surface of the cube shaded; in Figure 6, a face or the figure of a seated woman; in Figure 7, a duck (facing left) or a rabbit (facing right); in Figure 8, a white goblet or the black silhouettes of two faces in profile.

Name _____ Class _____ Date _____

Chapter 24   Regulation **Laboratory 24-3**

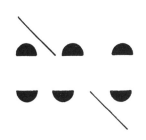

Figure 3

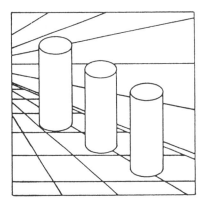

Figure 4

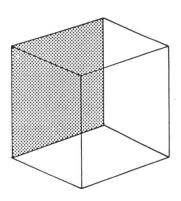

Figure 5

Figure 6

Figure 7

Figure 8

Biology   Copyright © by Globe Book Company   **123**

Analysis and Conclusions

1. Which of the figures you looked at in this laboratory are optical illusions?

   All are optical illusions.

2. Have you ever been in a car on a hot day and thought you saw water on the road ahead? Why is this an example of an optical illusion?

   because no water was actually on the road

3. What are some careers in which people would be concerned about optical illusions? Explain your answers.

   Accept all logical answers.

Extension

Use reference materials to help you create an optical illusion. Consider how the background of your drawing can affect the illusion. Draw your optical illusion in the space provided.

Name_____ Class_____ Date_____

Chapter 25   Reproduction and Development

Laboratory 25-1 What changes take place during childbirth?

Background Information

A human fetus develops inside the mother's body for 38 weeks. As the fetus develops, it goes through a series of changes. Finally, about nine months after fertilization, the fetus moves into a head-down position in preparation for birth. Contractions of the uterus, or labor, mark the beginning of the birth process.

Skills: organizing, sequencing, comparing

Objectives

In this laboratory, you will
- compare different stages in the birth process.

Prelab Preparation

Review Section 25-2 Human Reproduction.

Materials

paper
pencil

Procedure

1. Refer to Data Table 1, which lists several stages in the birth process.

2. Figures 1–4 show four of the stages in childbirth. Use these diagrams to help you complete Data Table 1. Put a plus sign (+) in the column if the event listed takes place during that stage; put a minus sign (–) in the column if the event does not take place during that stage.

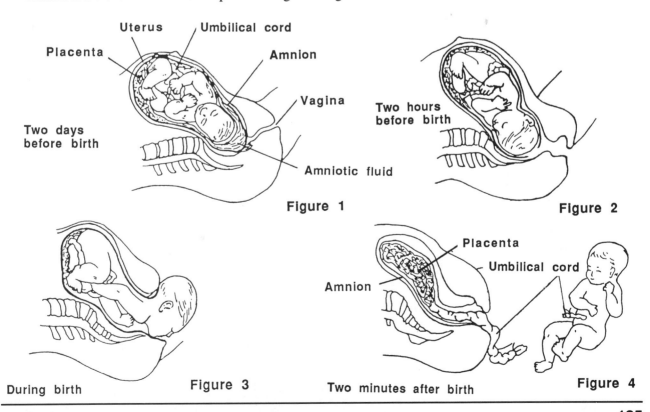

Figure 1

Figure 2

During birth   Figure 3

Two minutes after birth   Figure 4

Biology   Copyright © by Globe Book Company   **125**

Observations and Data

Data Table 1: Stages Before, During, and After Birth

Stages	Two Days Before Birth	Two Hours Before Birth	During Birth	Two Minutes After Birth
1. Vagina has opened.	−	+	+	+
2. Baby is attached to the mother.	+	+	+	−
3. Baby is attached to the umbilical cord.	+	+	+	−
4. Contractions are occurring.	−	+	+	−
5. Amnion has broken.	−	+	+	+
6. Baby is inside the amnion.	+	+	+	−
7. Baby is outside the mother's body.	−	−	−	+
8. Umbilical cord is attached to the placenta.	+	+	+	+
9. Baby is inside the uterus.	+	+	+	−
10. Placenta is attached to the uterus.	+	+	+	−
11. Amniotic fluid is in the amnion.	+	−	−	−
12. Placenta has been pushed out of the mother's body.	−	−	−	+

Name_____ Class_____ Date_____

Chapter 25 Reproduction and Development   Laboratory 25-1

1. Which of the events listed in Data Table 1 take place before birth? (List the numbers.)

   1, 2, 3, 4, 5, 6, 8, 9, 10, 11

2. Which events take place during birth?

   1, 2, 3, 4, 5, 6, 8, 9, 10

3. Which events take place after birth?

   1, 5, 7, 8, 12

Analysis and Conclusions

1. Organize the events listed in Data Table 1 in the correct sequence from the earliest to the latest.

   2, 3, 6, 8, 9, 10, 11, 1, 4, 5, 7, 12

2. Why is the placenta sometimes called the "afterbirth"?

   because it is pushed out of the uterus after the birth of the baby

Extension

Use reference materials to research a cesarean delivery. What is the origin of the term "cesarean"? Compare a cesarean delivery with a natural birth canal, or vaginal, delivery. Organize your information in a data table similar to Data Table 1. Use the space provided on the following page to record your findings.

Biology

Notes

Name _____ Class _____ Date _____

Chapter 26　　　　　　　　　　　　　　　　　　　　　　Nutrition

Laboratory 26-1 Which foods contain sugar and starch?

Background Information
Among the nutrients found in foods are carbohydrates, or sugars and starches. Carbohydrates provide fuel for your cells. Many different kinds of foods contain simple or complex sugars and starches, which are made up of many sugar molecules joined together.

Skills: observing, classifying, inferring

Objective
In this laboratory, you will
- test samples of different foods for the presence of sugar and starch.

Prelab Preparation
1. Review Section 26-1 Nutrients.

2. Review the proper technique for lighting and using a Bunsen burner.

3. Review the use of indicators to identify the presence of certain chemicals.

Materials

Lugol's solution	waxed paper	**Note to teacher:** In addition to the corn syrup and raw potato, have students bring in samples of various foods including cooked rice, white bread, egg whites, milk, and honey. You might also want to have students test samples of "junk" foods and "fast" foods.
Benedict's solution	water	
corn syrup	piece of raw potato	
food samples	Bunsen burner	
test tubes	test tube rack	
test tube holder	dropper	

Procedure
Part A: Testing for Sugar　

1. Add 3 mL of Benedict's solution to a clean, dry test tube. **Caution: Benedict's solution can irritate the skin. Wash with cold water immediately if any spills on your skin.**

2. Add 3 mL of corn syrup to the Benedict's solution. Swirl gently to mix the liquids.

3. Using a test tube holder, gently heat the solution in the test tube over a Bunsen burner. **Caution: Be careful when using a Bunsen burner.**

4. A color change to green, yellow, orange, or red indicates the presence of sugar. Record your observations in Data Table 1.

5. Test each of your food samples for the presence of sugar by repeating steps 1 to 4. If a food sample is solid, add 3 mL of water to the food in a test tube and swirl to mix before adding Benedict's solution. Record your observations in Data Table 1.

Biology　　　　　　　　　　　Copyright ©　　by Globe Book Company　　　　　　　　　129

Part B: Testing for Starch

1. Place a small piece of raw potato on a sheet of waxed paper.

2. Place three drops of Lugol's solution on the piece of potato.

3. A color change to blue-black indicates the presence of starch. Record your observations in Data Table 1.

4. Test each of your food samples for the presence of starch by repeating steps 1 to 3. If a food sample is liquid, place 3 mL of the liquid in a test tube and add three drops of Lugol's solution. Record your observations in Data Table 1.

Observations and Data

Data Table 1: Tests for Sugar and Starch

Food Sample	Color Change		Sugar Present? (Yes/No)	Starch Present? (Yes/No)
	Benedict's	Lugol's		
Corn syrup	red	none	yes	no
Potato	none	blue-black	no	yes
Cooked rice	none	blue-black	no	yes
Egg white	none	none	no	no
Milk	red	none	yes	no
Honey	red	none	yes	no
White bread	red	blue-black	yes	yes
Other:	Answers will vary.			

1. Which of the foods tested contained sugar?

   corn syrup, honey, bread, milk

   Other answers will vary depending on the foods tested.

2. Which of the foods tested contained starch?

   potato, rice, bread

   Other answers will vary depending on the foods tested.

Name _____ Class _____ Date _____

Chapter 26 Nutrition

Laboratory 26-1

---

3. Did any of the foods tested contain both sugar and starch? Which ones?

   bread

   Other answers will vary depending on the foods tested.

Analysis and Conclusions

1. What were the two controls used in this laboratory? What is the purpose of using a control?

   testing corn syrup for sugar and testing potato for starch

   A control is used for comparison.

2. Suppose two food samples are tested with Benedict's solution. In the first sample, the color changes to green. In the second sample, the color changes to red. How can you explain these different results?

   The food that changed color to red contained more sugar than the food that changed color to green.

Biology

Extension

Examine the nutrition information listed on the labels or packages of various foods. Record the following information in a data table: the amount of sugar, starch, or carbohydrates in each food and the percent of minimum daily requirements supplied by each food. List the foods in order from highest carbohydrate content to the lowest. What kind of food contained the greatest amount of carbohydrates? The least? Record your results in the space provided.

Name _____ Class _____ Date _____

Chapter 26 | Nutrition

Laboratory 26-2 What is the purpose of food additives?

Background Information
Any substance other than the main ingredient in a food is called an additive. Some additives prevent food from spoiling or separating. Others are used to add color or flavor, or to give foods a creamy texture. Nutrients such as vitamins and minerals also can be additives.

Skills: classifying, inferring, comparing

Objective
In this laboratory, you will
- identify additives in various foods and determine the purpose of each additive.

Prelab Preparation
1. What are preservatives?
   additives that protect foods against decay, spoiling, or discoloration

2. What are emulsifiers and stabilizers?
   additives that prevent the different ingredients in a food from separating

Materials
food labels
pencil
paper

Procedure
1. Read the list of ingredients on a food label or package.

2. Record each additive listed on the label in Data Table 1.

3. Refer to Figure 1 to find the purpose of each additive. Record this information in the "Purpose" column of Data Table 1.

4. Repeat steps 1 to 3 for as many different kinds of food as possible. Compare your results with your classmates' results.

Biology   Copyright © by Globe Book Company   133

Figure 1: Common Food Additives

Additives	Foods	Purpose
Agar	ice cream, frozen custard, sherbert	thickener
Ascorbic acid (vitamin C)	beverages, beverage mixes, fruit products	nutrient, preservative
BHA and BHT	bakery products, cereals, snack foods, fats and oils	preservatives
Calcium propionate	breads and other baked goods	preservative
Caramel	many types and varieties	coloring
Cellulose	breads, ice cream, sweets, diet foods	thickener, improved texture
Corn syrup	cereals, baked goods, candies, processed foods, processed meats	sweetener
Dextrose	cereals, baked goods, candies, processed foods, processed meats	sweetener
Gelatin	puddings, cheese spreads, cream cheese	thickener
Gum acacia (gum arabic)	soft drinks, imitation fruit juice drinks, ice cream	thickener, stabilizer
Hydrolyzed vegetable protein	processed meats, gravy and sauce mixes	flavoring
Iodine	salt	nutrient
Iron	grain products	nutrient
Pectin	jams, jellies, fruit products, frozen desserts	thickener, improved texture
Potassium sorbate	cheeses, syrups, cakes, beverages, mayonnaise, fruit products, margarine, processed meats	preservative
Riboflavin	flour, breads, cereals, rice, macaroni products	nutrient, coloring
Saccharin	special diet foods and beverages	sweetener
Saffron	many types and varieties	coloring
Sodium nitrate and sodium nitrite	cured meats, fish, poultry	preservatives
Sucrose	cereals, baked goods, candies, processed foods, processed meats	sweetener

Name _____ Class _____ Date _____

Chapter 26   Nutrition

Laboratory 26-2

Observations and Data

Data Table 1: Food Additives

Food	Additives	Purpose
	Answers will vary.	

1. Which additive was found most often in the foods you listed?

   Answers will vary.

2. What kind of additive (preservative, emulsifier, sweetener, etc.) was used most often in the foods you listed?

   Answers will vary.

3. Which of the foods you listed contained the most additives? Which contained the fewest?

   Answers will vary.

4. Did any of the foods you listed contain no additives? Which ones?

   Answers will vary.

Biology   Copyright © by Globe Book Company

Analysis and Conclusions

1. Why are substances other than the main ingredient of a food called additives?
   because they are "added to" the main ingredient

2. Why do you think many foods contain "nutritional supplements"?
   Accept all logical answers.

3. Why is baking powder considered an additive in bread?
   because it is not one of the main ingredients

4. Why do you think sugar is added to so many different kinds of foods?
   to sweeten the food, mask bitterness, change the flavor, etc. Accept all logical answers.

Extension

Use reference materials to research the different sugars and other sweeteners used in foods. In a data table, list both the natural and artificial sweeteners that can be added to food instead of sugar. What are the advantages and disadvantages of using natural or artificial sweeteners? Use the space provided to record your findings.

Name _____ Class _____ Date _____

Chapter 27   Diseases and Disorders

Laboratory 27-1 What conditions can prevent milk from spoiling?

Background Information

As bacteria grow, they cause physical and chemical changes in food. These changes cause the food to spoil. To prevent food from spoiling, the growth of bacteria must be stopped or slowed. Bacterial growth can be slowed by controlling the conditions under which food is prepared and stored.

Skills: observing, comparing, predicting

Objectives

In this laboratory, you will
- compare the rate of spoilage in samples of milk stored under different conditions.
- predict which conditions will prevent milk from spoiling.

Prelab Preparation

1. Review Section 27-1 Infectious Diseases.

2. Review the procedure for using a compound microscope.

Materials

compound microscope	microscope slides	cover slips
dropper	marking pencil	3 test tubes
test tube rack	milk	graduated cylinder

Procedure

1. Place one drop of milk on a clean, dry microscope slide. Carefully lower a cover slip over the drop of milk. Be sure to avoid trapping air bubbles in the milk.

2. Observe the drop of milk under the low power of a microscope. Then switch to high power.

3. Record the appearance of the milk in Data Table 1 under "Day 1." Include color, odor, texture, and so forth. **Caution: Do not taste the milk.**

4. Label three test tubes A, B, and C.

5. Add 3 mL of milk to each of the test tubes.

6. Place test tube A in a refrigerator, test tube B in a dark place, and test tube C in a window where it will receive direct light.

7. Predict which sample of milk — A, B, or C — will be the first to spoil and which will be the last. Record your predictions in Data Table 2.

8. Observe your milk samples every day for the next four days. Repeat steps 1, 2, and 3 for each sample.

9. After five days, complete Data Table 2 by recording the actual results.

Biology   Copyright ©   by Globe Book Company

Observations and Data

Data Table 1: Appearance of Milk

Test Tube	Day 1	Day 2	Day 3	Day 4	Day 5
A (cold)	milk should appear smooth, white, and milky.		Answers will vary.		
B (dark)					
C (light)					

Data Table 2: Milk Spoilage

Spoilage	Prediction	Actual
First to spoil	Answers will vary.	test tube C
Last to spoil		test tube A

1. Which milk sample was the first to spoil? Which sample was last?
   C, A

2. Were your predictions correct?
   Answers will vary.

Analysis and Conclusions

1. What conditions prevented the milk from spoiling? Use your data to support your answers.
   Lowering the temperature of the milk by placing it in a refrigerator prevented the milk from spoiling.

2. Why did these conditions prevent the milk from spoiling?
   Lowering the temperature slowed the growth of bacteria in the milk.

Extension

Do you think the kind of milk you used in this laboratory had any effect on the rate of spoilage? Repeat this laboratory using different kinds of milk, such as skimmed milk, lowfat milk, buttermilk, powdered milk, and condensed milk. Record your observations in a data table. Use reference materials to find out what causes the physical and chemical changes that take place in milk as it spoils.

Name _____ Class _____ Date _____

Chapter 27   Diseases and Disorders

Laboratory 27-2 What are some ways in which diseases can be spread?

Background Information

Infectious diseases are caused by bacteria and viruses. Bacteria and viruses cannot move from person to person on their own. Instead, they are transmitted by contact with an infected person or a contaminated object.

Skills: collecting and recording data, comparing, predicting

Objective

In this laboratory, you will
- observe some ways in which bacteria can be transmitted from place to place and from person to person.

Prelab Preparation

1. Review Section 27-1 Infectious Diseases.

2. What is nutrient agar? Why is it used as a growth medium for bacteria?
   Nutrient agar is made from brown algae; it contains nutrients necessary for bacterial growth.

Materials

2 Petri dishes	cotton swabs	nutrient agar
hand lens	marking pencil	coin

Procedure

1. Fill a Petri dish with nutrient agar. Place the lid on the Petri dish and label the dish A. This will be your control dish.

2. Turn a second Petri dish upside down. With a marking pencil, divide the dish into three sections. Label the sections 1, 2, and 3 as shown in Figure 1.

3. Turn the dish right-side up. Label the lid of this dish B. This will be your experimental dish.

4. Fill dish B with nutrient agar. Place the lid on the dish and let the agar solidify.

5. Shake hands with your partner.

6. Open dish B and carefully touch the surface of the agar in section 1 with the tips of two fingers.

7. Place a coin on the surface of the agar in section 2, and then remove it.

8. Touch a cotton swab to your tongue.

9. Gently rub the cotton swab back and forth over the surface of the agar in section 3.

10. Replace the lid on Petri dish B. Tape the lid shut.

11. Place both Petri dishes in a warm place where they will not be disturbed for two days.

12. Refer to Data Table 1. Predict the order in which bacterial growth will appear in Petri dish B.

13. Observe both Petri dishes after two days. Draw what you see in Plates 1 and 2 and record your actual results in Data Table 1.

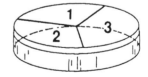

Figure 1

Biology   Copyright ©   by Globe Book Company   139

Observations and Data

Data Table I: Bacterial Growth

Section	Predicted Growth	Actual Growth
1 (fingertips)	Answers will vary.	
2 (coin)		
3 (moist swab)		

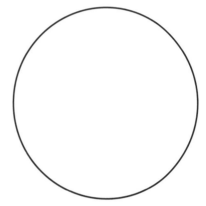

Plate 1: Petri dish A

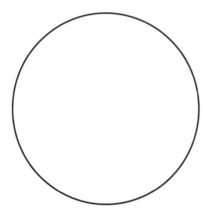

Plate 2: Petri dish B

1. Compare the amount of bacterial growth in Petri dish A with the growth in Petri dish B.
   Answers will vary. Students should observe more growth in dish B than in dish A.

2. Were your predictions about bacterial growth in dish B correct?
   Answers will vary.

Analysis and Conclusions

1. What are some ways in which disease-causing bacteria can be spread? Use your data to support your answer.
   through direct contact with an infected person or a contaminated object

2. What are three ways to prevent the transmission of infectious diseases?
   Answers will vary. Students should mention avoiding contact with infected people, using antibiotics or disinfectants to kill bacteria, etc. Accept all logical responses.

Extension

How would washing your hands with soap and water, soaking the coin in a disinfectant, and rinsing your mouth with mouthwash before touching the agar affect the growth of bacteria? To answer this question, repeat the laboratory following these procedures. Record your observations in a data table.

Name _____ Class _____ Date _____

Chapter 28 — Drugs, Alcohol, and Tobacco

Laboratory 28-1 Which antacids are most effective in neutralizing acid?

Background Information

Antacids are over-the-counter drugs. They are used to neutralize stomach acids that can cause pain or discomfort. Not all antacids are equally effective in neutralizing acid, however. One way to test the effectiveness of an antacid is to measure the amount of acid it neutralizes.

Skills: inferring, comparing, recording data

Objective

In this laboratory, you will
- test three different antacids to find out which neutralizes the most acid.

Prelab Preparation

1. Review Section 28-1 Types of Drugs.
2. Review the function of an indicator.

Materials

sodium bicarbonate solution	Congo red solution	3 antacid solutions
hydrochloric acid	stirring rod	8 test tubes
test tube rack	dropper	marking pencil
graduated cylinder		

Procedure

1. Label two test tubes A and B.

2. Pour 20 mL of sodium bicarbonate solution into each test tube.

3. Carefully add three drops of Congo red solution to each test tube. Use a stirring rod to mix the liquids. **Caution: Congo red is a dye that can stain your skin and clothing.**

4. Slowly add hydrochloric acid, one drop at a time, to test tube A. Stir the solution after each drop. **Caution: Be very careful when using acids. If any acid spills on your skin or clothing, wash it off immediately and tell your teacher.**

5. A color change from red to purple indicates that the acid has been neutralized. Count the number of drops of acid needed to cause the sodium bicarbonate solution to change color from red to purple. Record this number in Data Table 1.

6. Label six test tubes C through H.

7. Add 20 mL of an antacid solution to test tubes C and D.

8. Add three drops of Congo red solution to each test tube.

9. Slowly add hydrochloric acid one drop at a time to test tube C. Stir after each drop and count the number of drops needed to change the color of the solution from red to purple. Record the number of drops in Data Table 1.

10. Repeat steps 7–9 with test tubes E and F, using a different antacid solution. Be sure to add acid to test tube E only.

11. Repeat steps 7–9 with test tubes G and H, using a third antacid solution. Be sure to add acid to test tube G only.

12. Discard the liquids in the test tubes, following your teacher's directions. Clean and rinse the test tubes thoroughly.

Biology   Copyright ©   by Globe Book Company   141

Observations and Data

Data Table I: Neutralization of Acid

Antacid	Drops of Acid Added
Test tube A (sodium bicarbonate)	
Test tube C (brand name:)	Answers will vary.
Test tube E (brand name:)	
Test tube G (brand name:)	

1. Which antacid took the greatest number of drops of acid before changing color? The fewest?
   Answers will vary.

2. How did you know when the antacid had neutralized as much acid as possible?
   The color of the solution changed from red to purple.

Analysis and Conclusions

1. What were the controls in this laboratory?
   test tubes B, D, F, and H, to which acid was not added

2. What was the purpose of these controls?
   to compare the color changes that took place in the other test tubes

3. Which antacid was most effective? Use your data to support your answer.
   Answers will vary. The antacid that took the greatest number of drops of acid was most effective.

4. List the antacids you tested in order from most effective to least effective.
   Answers will vary.

5. Compare your results with those of your classmates. How can you explain any differences in the results?
   Answers will vary. Accept all logical responses.

Extension

Read the label for each brand of antacid you tested. Add the name of the main chemical ingredient listed for each antacid to Data Table 1. According to your data, which ingredient is most effective? Compare the cost of each antacid with its effectiveness. Is the most expensive brand also the most effective? Do you think using a different acid would change the effectiveness of the antacids tested? To answer this question, repeat the laboratory using an acid solution of 35 mL of vinegar in 500 mL of water.

Name _____ Class _____ Date _____

Chapter 28 — Drugs, Alcohol, and Tobacco

Laboratory 28-2 How do alcohol and tobacco affect the germination of seeds?

Background Information

Ethyl alcohol, the alcohol in alcoholic drinks, is a commonly used and abused drug. Alcohol is a depressant that slows down the central nervous system. The tobacco in cigarettes and cigars comes from the dried leaves of tobacco plants. Tobacco contains the drug nicotine as well as many other harmful substances.

Skills: inferring, comparing, measuring

Objectives

In this laboratory, you will
- observe the effects of alcohol and tobacco on the germination of lima bean seeds.

Prelab Preparation

1. Review Section 28-2 Alcohol and Section 28-3 Tobacco.

2. Review the procedure for germinating seeds.

3. Review how to find an average.

Note to teacher: You may wish to prepare the solutions and soak the lima beans for 24 hours before students begin this laboratory.

Materials

lima beans	water	ethanol
cigarette tobacco	paper towels	3 baby food jars
metric ruler	hand lens	marking pencil

Procedure

1. Soak four lima beans in water, four in a solution of ethanol and water, and four in a solution of cigarette tobacco and water. Allow the lima beans to soak for 24 hours.

2. Label three baby food jars 1, 2, and 3.

3. Fold three paper towels and place one towel in each jar. Moisten each towel with water.

4. Place the four lima beans that were soaked in water into jar 1. Be sure the beans are between the moist paper towel and the side of the jar.

5. Place the four lima beans that were soaked in ethanol and water into jar 2, and the four that were soaked in tobacco and water into jar 3.

6. Put the three jars in a dark, warm place where they will not be disturbed.

7. Observe each jar every day for five days. Count the number of seeds in each jar that germinated. Record your observations in Data Table 1.

8. Measure the amount of new growth on each bean seed every day for five days. Calculate the average new growth for each seed that germinated. Record your observations in Data Table 2.

Biology   Copyright © by Globe Book Company

Observations and Data

Data Table 1: Seed Germination

Jar	Number of Seeds Germinated				
	Day 1	Day 2	Day 3	Day 4	Day 5
1 (water)			Answers will		
2 (alcohol)			vary.		
3 (tobacco)					

Data Table 2: Seed Growth

Jar	Average New Growth (mm)				
	Day 1	Day 2	Day 3	Day 4	Day 5
1 (water)			Answers will		
2 (alcohol)			vary.		
3 (tobacco)					

1. How did the number of germinated lima bean seeds soaked in water compare with the number soaked in the alcohol solution and in the tobacco solution?
   More of the seeds that were soaked in water germinated than those that were soaked in either the alcohol solution or the tobacco solution.

2. Which lima beans showed the most average growth after five days?
   The lima beans that were soaked in water

3. Which lima beans showed the least average growth after five days?
   Answers will vary.

Analysis and Conclusions

1. What effect did soaking lima beans in alcohol have on the germination of the beans?
   Not all of the lima beans germinated and those that did germinate had a low average growth.

2. What effect did soaking lima beans in tobacco and water have on the germination of the beans?
   Not all of the lima beans germinated and those that did germinate had a low average growth.

Extension

Did the kind of seeds used in this laboratory have any effect on the results? To answer this question, repeat the laboratory using different kinds of seeds. What effect does the kind of tobacco used have on the germination of seeds? Repeat this laboratory using different kinds of tobacco, such as pipe tobacco, cigar tobacco, and smokeless tobacco. Record your observations in a data table.

Name _____ Class _____ Date _____

Chapter 28 — Drugs, Alcohol, Tobacco

Laboratory 28-3 What are habits?

Background Information

When a learned act becomes automatic, it is called a habit. Some habits, such as tying a shoelace or writing your name, are helpful. Other habits, such as smoking or biting your fingernails, are not helpful. In this Laboratory, you will investigate some of your habits. You will try to break an old habit and form a new one.

Skills: observing, applying

Objective

In this laboratory, you will
- observe and identify the writing habits of yourself and others.

Materials

pencil   stopwatch or watch with second hand   paper

Procedure

1. Work with a partner. With the hand that you usually use, write your name as many times as you can in one minute. Have your partner time you. Record the results in Table 1.

2. Now use your other hand and see how many times you can write your name in one minute. Try this two more times, having your partner time you each time. Record your results in Table 1.

3. Have your partner write the digits 0123456789 several times. Observe your partner carefully as he or she writes.

4. Did your partner begin to write each digit at the top or bottom of the digit? Record your observations in Table 2.

5. Now write the digits yourself and observe where you begin to write each digit. Record your observations in Table 2.

6. Dotting i's and crossing t's are a writing habit. Test the strength of this habit. Have your partner read aloud the passage below. As he or she reads, write out the passage. Write as fast as you can, but do not cross your t's or dot your i's.

    Little Tommy tried to tie a tiny toy tin box. He tipped it over and all the little toys inside fell out. Tommy tried not to talk about the accident.

7. After you are finished, count the number of t's that you crossed in the passage. Also count the number of i's that you dotted. Write your results in Table 3.

Biology   Copyright © by Globe Book Company

Observations

Table 1 Writing Your Name

Hand	Times per Minute
Usual Hand	Answers will vary.
Other Hand: Trial 1	
Trial 2	
Trial 3	

Table 2 Writing Digits

Person	Where Began Writing
your partner	Answers will vary.
you	

Table 3 Crossing t's and Dotting i's

Letter	Number in Passage	Number Crossed or Dotted
t	24	Answers will vary.
i	12	

Analysis and Conclusions

1. Which one of your hands do you have the habit of using to write your name?
   Answers will vary.

2. What evidence do you have that your other hand could learn this habit?
   Answers will vary. Each trial for writing with the other hand improved.

3. Where do you think that you formed the habit of writing your digits starting from the top?
   In school

4. What evidence can you present to show that dotting i's and crossing t's is a habit?
   Even when students tried not to dot i's and cross t's, they did it anyway.

Extension

Habits can be broken. Here are five steps that can be followed to break a bad habit: 1. Recognize the habit; 2. want to break it; 3. substitute a good habit for the bad habit; 4. stop the bad habit all at once; 5. give yourself a reward for breaking the habit. On a separate sheet of paper, write out a five-step plan to break a habit that you do not want.

Name_____ Class_____ Date_____

Chapter 29   Fundamentals of Genetics

Laboratory 29-1 How can a test cross help to determine genotype?

Background Information

Some traits are dominant over other traits. An organism can have two dominant genes for a trait, two recessive genes for the trait, or be a hybrid with one dominant and one recessive gene for the trait. An organism that has pure dominant genes has the same phenotype, or appearance, as a hybrid. If an organism's genotype is unknown, geneticists can perform a test cross. In a test cross, an organism with an unknown genotype is crossed with an organism that is pure recessive for the trait.

Skills: modeling, inferring

Objectives

In this laboratory, you will
- use a model to simulate a test cross between two guinea pigs.
- infer the genotype of the unknown guinea pig based on the results of your test cross.

Prelab Preparation

Review Section 29-2 Probability and Genetics.

Materials

two bags of dried beans

Note to teacher: One bag should contain 10 dark beans and 10 white beans to represent the Bb hybrid. The other bag should contain 20 white beans to represent the recessive bb genotype. Label the Bb bag "Black fur" and the bb bag "White fur."

Procedure

1. Two bags of dried beans labeled "Black fur" and "White fur" will represent the genes of two guinea pigs that are being test crossed.

2. Without looking inside, take one bean out of each bag. The two beans represent the genotype of the offspring in the test cross. A dark bean represents a B gene and a white bean represents a b gene. Record the genotype and phenotype of the offspring in Data Table 1.

3. Return each bean to its correct bag. Shake each bag and take one bean from each bag. Record the results of this second test cross in Data Table 1.

4. Repeat steps 2 and 3 until you have recorded the results of 20 test crosses.

Biology

Observations and Data

Data Table 1: Test Crosses of Guinea Pigs

Test Cross	Color of Beans	Genotype	Phenotype (Fur Color)
1		Answers will vary.	
2			
3			
4			
5			
6			
7			
8			
9			
10			
11			
12			
13			
14			
15			
16			
17			
18			
19			
20			

Name _____ Class _____ Date _____

Chapter 29-1 Fundamentals of Genetics   Laboratory 29-1

1. How many of the offspring had black fur?
   approximately 10

2. How many had white fur?
   approximately 10

Analysis and Conclusions

1. What was the genotype of the offspring with black fur? Use your data to support your answer.

   Bb

2. What was the genotype of the offspring with white fur?

   bb

3. To produce these genotypes, what was the genotype of the unknown guinea pig with black fur—BB or Bb? How do you know?

   Bb, because crossing a BB guinea pig with a bb guinea pig would have produced all hybrid black (Bb)

   offspring

4. Use the space provided to draw a Punnett square representing the test cross you performed in this laboratory.

                   BB x bb

	B	b
b	Bb	bb
b	Bb	bb

Biology — Copyright © by Globe Book Company

Extension

What would the genotype of the unknown guinea pig have to be in order to produce all black offspring in a series of test crosses with a white guinea pig? Put the correct beans in each bag and repeat this laboratory to test your prediction. Was your prediction correct?

Name _____ Class _____ Date _____

Chapter 29 Fundamentals of Genetics

Laboratory 29-2 How can the results of monohybrid and dihybrid crosses be predicted?

Background Information

A cross involving only one trait is called a monohybrid cross; a cross involving two traits is a dihybrid cross. Both of these crosses can be shown using Punnett squares. A completed Punnett square can be used to predict the results of the crosses.

Skills: predicting, analyzing

Objective

- predict the results of a monohybrid cross and a dihybrid cross.

Prelab Preparation

Review Section 29–3 Predicting the Results of Genetic Crosses.

Materials

paper
pencil

Procedure

Part A: Monohybrid Cross

1. Coleus, a common houseplant, can have leaves with normal, shallow edges or leaves that are deeply cut or lobed. The trait for deeply lobed leaves is dominant over the trait for shallow-edged leaves. The Punnett square in Data Table 1 shows a cross between a coleus with deeply lobed leaves (DD) and a normal coleus (dd). Complete the Punnett square to predict the possible genotypes of offspring in the F_1 generation.

2. The Punnett square in Data Table 2 shows a cross between two individuals in the F_1 generation. Complete the Punnett square to predict the possible genotypes of offspring in the F_2 generation.

Part B: Dihybrid Cross

1. In corn, purple color and starchy kernels (PPSS) are dominant over yellow color and sweet kernels (ppss). The Punnett square in Data Table 3 shows a cross between purple, starchy corn and yellow, sweet corn. Complete the Punnett square to predict the possible genotypes of offspring in the F_1 generation.

2. The Punnett square in Data Table 4 shows a cross between two individuals in the F_1 generation. Complete the Punnett square to predict the possible genotypes of offspring in the F_2 generation.

3. Complete Data Table 5.

Observations and Data

Data Table 1: DD x dd

	D	D
d	Dd	Dd
d	Dd	Dd

Biology   Copyright © by Globe Book Company

Data Table 2: F₁ Generation (Dd x Dd)

	D	d
D	DD	Dd
d	Dd	dd

Data Table 3: PPSS x ppss

	PS	PS
ps	PpSs	PpSs
ps	PpSs	PpSs

Data Table 4: F₁ Generation (PpSs x PpSs)

	PS	Ps	pS	ps
PS	PPSS	PPSs	PpSS	PpSs
Ps	PPSs	PPss	PpSs	Ppss
pS	PpSS	PpSs	ppSS	ppSs
ps	PpSs	Ppss	ppSs	ppss

Data Table 5: Genotype and Phenotype Frequency

Genotype	Genotype Frequency	Phenotype	Phenotype Frequency
PPSS	1	purple, starchy	9
PPSs	2	purple, starchy	
PpSS	2	purple, starchy	
PpSs	4	purple, starchy	
PPss	1	purple, sweet	3
Ppss	2	purple, sweet	
ppSS	1	yellow, starchy	3
ppSs	2	yellow, starchy	
ppss	1	yellow, sweet	1

Chapter 29 Fundamentals of Genetics

Laboratory 29-2

Analysis and Conclusions

1. Use Data Table 1 to predict the phenotypes of the offspring in the F_1 generation of the monohybrid cross.
   All the offspring would have deeply lobed leaves.

2. a. Use Data Table 2 to predict the genotype ratio of the offspring in the F_2 generation.
   1 (DD) : 2 (Dd) : 1 (dd)

   b. Predict the phenotype ratio (deeply lobed to normal leaves) of the offspring in the F_2 generation.
   3:1

3. Use Data Table 4 to predict the phenotypes of the offspring in the F_1 generation of the dihybrid cross.
   All of the offspring would have purple, starchy kernels.

4. a. How many different phenotypes are produced in the F_2 generation of a dihybrid cross? Use your data to support your answer.
   four

   b. What is the characteristic phenotype ratio in the dihybrid cross?
   9:3:3:1

Extension

Obtain an ear of dihybrid corn. Notice the difference between the purple kernels and the yellow kernels. The starchy kernels will appear plump and the sweet kernels will appear wrinkled. Count the number of each type of kernel on the ear of corn. Do your actual results come close to your predicted ratio?

Notes

Name _____ Class _____ Date _____

Chapter 30 — Modern Genetics

Laboratory 30-1 What percentage of students in your class have inherited some common genetic traits?

Background Information
Many common physical characteristics are determined by dominant or recessive genes. For example, tongue rolling is the ability to roll the tongue into a U–shape. This ability to roll the tongue is a dominant trait. Other genetic traits include free ear lobes, a widow's peak, a straight thumb, and right–handedness.

Skills: observing, recording data, analyzing

Objective
In this laboratory, you will
- survey students in your class to calculate how frequently common genetic traits occur.

Prelab Preparation
1. Review Section 30–3 Human Genetics.

2. Review the procedure for calculating a percentage.

3. Review the descriptions of the following genetic traits.

    a. Tongue rolling: The ability to roll up the edges of the tongue so that the tongue forms a U–shape.
    b. Free ear lobes: The lobe of the ear hangs freely below the point of attachment to the head.
    c. Widow's peak: The midpoint of the hairline across the forehead forms a distinct peak.
    d. Straight thumb: When the thumb is extended directly up from the palm, the top part and the bottom part of the thumb form a straight line.
    e. Left thumb over right: When the hands are folded naturally, the left thumb crosses over the right thumb.
    f. Left arm over right: When the arms are crossed naturally in front of the chest, the left arm crosses over the right arm.
    g. Right–handedness: When performing common tasks, such as picking up a pencil, the right hand is used more often than the left hand.

Materials
paper
pencil

Note to teacher: You may wish to have students survey a certain number of their classmates and then collect and record their data in a class data table.

Procedure
1. Count the number of students in your class. Record this number in Data Table 1.

2. Ask each student in your class to roll his/her tongue. Record the number of males and females who are tongue rollers in Data Table 1.

3. Record the number of males and females who have free ear lobes in Data Table 1.

4. Record the number of males and females who have a widow's peak in Data Table 1.

5. Ask each student to hold his/her thumb straight up. Record the number of males and females who have a straight thumb.

Biology   Copyright © by Globe Book Company   155

6. Ask each student to fold his/her hands naturally. Observe which thumb is on top. Record the number of males and females who cross the left thumb over the right.

7. Ask each student to cross his/her arms. Observe which arm is on top. Record the number of males and females who cross the left arm over the right.

8. Ask each student to pick up a pencil. Record the number of males and females who use the right hand.

9. Complete Data Table 2 by calculating the percentage of males and females who have each of the genetic traits listed.

Observations and Data

Trait	Number of Students	Number of Males	Number of Females	Percentage of Males	Percentage of Females
Tongue rolling			Answers will		
Free ear lobes			vary.		
Widow's peak					
Straight thumb					
Left thumb					
Left arm					
Right–handed					

Analysis and Conclusions

1. Are the traits listed in Data Table 1 dominant or recessive traits? How do you know?
   The traits listed are dominant. More than 50% of the students surveyed should have these traits.

2. Do you think any of the traits listed are sex–linked traits? Use your data to support your answer.
   Answers will vary based on students' data. Accept all logical responses.

Extension

Design an experiment to survey 50 people for the traits listed in this laboratory. Record the percentages in a data table. Use your data to estimate the percentage of males and females in a group of 500 people who would have each trait.

Name_____ Class_____ Date_____

Chapter 30 — Modern Genetics

Laboratory 30-2 How do normal red blood cells and sickled cells compare?

Background Information

Sickle–cell anemia is a genetic disorder that is caused by a recessive gene. A person who has this disorder inherited a pair of recessive genes (ss). A person who is hybrid for this disorder inherited one recessive, defective gene and one dominant, normal gene (Ss). Such a person is healthy, but carries the sickle–cell trait.

Skills: observing, comparing

Objectives

In this laboratory, you will
- compare normal red blood cells with red blood cells showing sickle–cell anemia and sickle–cell trait.
- use Punnett squares to predict how sickle–cell anemia is inherited.

Prelab Preparation

1. Review Section 30–4 Genetic Disorders.

2. Review the procedure for using a compound microscope.

Materials

compound microscope
prepared slides of normal human blood, blood showing sickle–cell trait and sickle–cell anemia

Procedure

Part A: Observing Normal and Sickled Blood Cells

1. Place a prepared slide of normal human red blood cells on the stage of a compound microscope.

2. Examine the slide under the low power of the microscope. Then switch to high power. Draw what you see in Plate 1.

3. Examine a prepared slide of red blood cells from a person with sickle–cell trait. Draw what you see in Plate 2.

4. Examine a prepared slide of red blood cells from a person with sickle–cell anemia. Draw what you see in Plate 3.

Part B: Predicting the Inheritance of Sickle–Cell Anemia

1. Predict the possible genotypes of the offspring of a pure dominant (SS) parent and a hybrid (Ss) parent. Record your predictions by completing the Punnett square in Data Table 1.

2. Predict the possible genotypes of the offspring of two hybrid (Ss) parents. Record your predictions by completing the Punnett square in Data Table 2.

3. Predict the possible genotypes of the offspring of a pure dominant (SS) parent and a pure recessive (ss) parent. Record your predictions by completing the Punnett square in Data Table 3.

4. Predict the possible genotypes of the offspring of a hybrid (Ss) parent and a pure recessive (ss) parent. Record your predictions by completing the Punnett square in Data Table 4.

Observations and Data

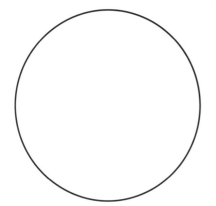

Plate 1:   Normal

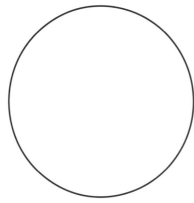

Plate 2:   Sickle–cell Trait

Plate 3:   Sickle–cell anemia

Data Table 1:   SS x Ss

	S	S
S	SS	SS
s	Ss	Ss

Data Table 2:   Ss x Ss

	S	s
S	SS	Ss
s	Ss	ss

Data Table 3:   SS x ss

	S	S
s	Ss	Ss
s	Ss	Ss

Data Table 4:   Ss x ss

	S	s
s	Ss	ss
s	Ss	ss

Name _____ Class _____ Date _____

Chapter 30   Modern Genetics   Laboratory 30-2

1. Compare the shape of normal red blood cells with the shape of sickled cells.
   Normal red blood cells look like flat disks; sickle cells look like irregular, half–moon shapes.

   Accept all logical answers.

2. Compare the number of normal and sickled cells in a person with sickle–cell trait and a person with sickle–cell anemia.
   In a person with sickle–cell trait, about half the cells are sickled. In a person with sickle–cell anemia, all of the cells are sickled.

Analysis and Conclusions

1. a. Refer to Data Table 1. How many of the children of these parents could have sickle–cell anemia?
   none

   b. Refer to Data Table 2. How many of these children could have sickle–cell anemia?
   one–fourth

   c. Refer to Data Table 3. How many of these children could have sickle-cell anemia?
   none

   d. Refer to Data Table 4. How many of these children could have sickle–cell anemia?
   one–half

2. Refer to Data Table 2. Predict the chances of these parents having a normal child, a child with sickle–cell trait, and a child with sickle cell anemia.

   normal: **25% (1/4)**

   sickle–cell trait: **50% (1/2)**

   sickle–cell anemia: **25% (1/4)**

3. Why do you think a person with sickle–cell trait is called a carrier?
   because the person has, or carries, one recessive, defective gene that can be passed on to offspring

Extension

Huntington's disease is a rare genetic disease that affects the brain and is always fatal. Scientists have now developed a test that allows a person to find out if he or she has Huntington's disease before any symptoms of the disease appear. Use reference materials to research Huntington's disease. Record your findings in a written report. If you knew that Huntington's disease occurred in your family, would you take the test? Give reasons for your answer in your report.

Name _____ Class _____ Date _____

Chapter 30 | Modern Genetics

Laboratory 30-3 How are sex–linked traits inherited?

Background Information

Hemophilia is a sex-linked trait; the genes for hemophilia are located on the X chromosomes. The normal, dominate gene is X^H. The recessive, defective gene is X^h. A female can have the genotype $X^H X^H$, $X^H X^h$, or $X^h X^h$. A male can have the genotype $X^H Y$ or $X^h Y$.

Skills: modeling, inferring, analyzing

Objective

In this laboratory, you will
- use a model to show how the sex–linked trait hemophilia is inherited.

Prelab Preparation

Review Section 30–3 Human Genetics.

Materials

4 coins
tape
pencil

Procedure

Part A: $X^H X^h$ Mother and $X^H Y$ Father

1. Place a piece of tape on both sides of two coins. Label the coins as shown in Figure 1. These coins will represent the genes of the mother and father.

2. Shake the coins in your hands and drop the coins on your work surface. The letters that appear on the coins represent the genotype of the offspring.

3. Record the genotype of the offspring in Data Table 1 by making a slash mark (/) in the proper column under the heading "Offspring."

4. Repeat steps 2 and 3 a total of 50 times.

5. Complete Data Table 1 by adding up the number of marks for each genotype and recording these numbers under the heading "Total."

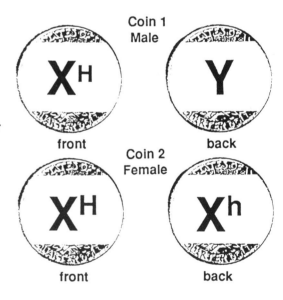

Figure 1

Part B: $X^H X^h$ Mother and $X^h Y$ Father

1. Place a piece of tape on both sides of two coins. Label the coins as shown in Figure 2. These coins will represent the genes of the mother and father.

2. Shake the coins in your hands and drop the coins on your work surface. The letters that appear on the coins represent the genotype of the offspring.

3. Record the genotype of the offspring in Data Table 2 by making a slash mark (/) in the proper column under the heading "Offspring."

4. Repeat steps 2 and 3 a total of 50 times.

5. Complete Data Table 2 by adding up the number of marks for each genotype and recording these numbers under the heading "Total"

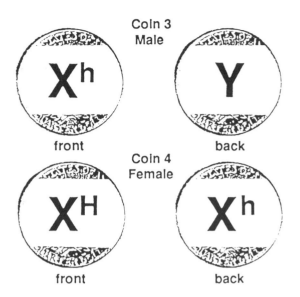

Figure 2

Observations and Data

Data Table 1: $X^H X^h$ Mother and $X^H Y$ Father

Genotype	Phenotype	Offspring	Total
$X^H X^H$	normal female	Answers will vary.	
$X^H X^h$	carrier female		
$X^h X^h$	hemophiliac female		
$X^H Y$	normal male		
$X^h Y$	hemophiliac male		

Data Table 2: $X^H X^h$ Mother and $X^h Y$ Father

Genotype	Phenotype	Offspring	Total
$X^H X^H$	normal female	Answers will vary.	
$X^H X^h$	carrier female		
$X^h X^h$	hemophiliac female		
$X^H Y$	normal male		
$X^h Y$	hemophiliac male		

Name _____ Class _____ Date _____

Chapter 30 Modern Genetics   Laboratory 30-3

Analysis and Conclusions

1. Refer to Data Table 1. Could these parents have a hemophiliac son? Could they have a hemophiliac daughter?
   yes; no

2. Refer to Data Table 2. Could these parents have a hemophiliac son? Could they have a hemophiliac daughter?
   yes; yes

3. In order for a female to have hemophilia, what genes must she inherit from her parents? Explain your answer.
   Because a female has two X chromosomes, she must inherit two recessive, defective genes from her parents in order to have hemophilia.

4. Can a male be a carrier of hemophilia? How do you know?
   no, because a male has only one X chromosome; therefore, if he carries the gene for hemophilia, he will have hemophilia

5. Why is hemophilia called a sex–linked trait?
   because the gene for hemophilia is carried on the X chromosome

Extension

Repeat this laboratory to predict the inheritance of color blindness. Also predict the results of a cross between a colorblind female and a normal male. Use the space provided on the next page to record your results.

Biology   Copyright ©   by Globe Book Company   163

Notes

Name _____ Class _____ Date _____

Chapter 31   Applied Genetics

Laboratory 31-1 How can a karyotype be used to identify human genetic disorders?

Background Information

A karyotype is a photograph of the chromosomes in a human body cell. When the photograph is enlarged, a genetics laboratory technician can cut out the individual chromosomes, arrange them in matching pairs, and organize the chromosome pairs on a karyotype chart. By studying this chart, it is possible to recognize chromosome abnormalities that could result in genetic disorders.

Skills: observing, comparing and contrasting, organizing data

Objective

In this laboratory, you will
- use a human karyotype chart to identify a genetic disorder.

Prelab Preparation

What is Down's syndrome? What causes Down's syndrome?

Down's syndrome is a genetic disorder in which a person has three chromosomes in pair 21 instead of two.

It results when chromosome pairs do not separate correctly during meiosis.

Materials

scissors
tape

Procedure

1. Refer to Figure 1, which is a picture of human chromosomes. Use scissors to carefully cut out each chromosome shown in Figure 1. Cut a square or rectangle shape around each chromosome. **Caution: Be careful when using scissors.**

2. Arrange the chromosomes in pairs according to size. Then place the matched pairs of chromosomes on the human karyotype chart in Figure 2. Place the longest pair of chromosomes above line number 1, the next longest pair above number 2, and so on.

3. The last pair of chromosomes are the sex chromosomes. **Note:** In males, the sex chromosomes are not a matched pair. Place the last two chromosomes above the line labeled "sex chromosomes."

4. When you are sure you have all the chromosomes in their proper place on the chart, tape them to the chart.

Observations and Data

1. How many chromosomes are shown on the completed karyotype chart?

   47

2. Does this chart represent a male or a female? How do you know?

   male, because there is one X and one Y chromosome

Analysis and Conclusions

1. Does your completed chart show a normal human karyotype? How do you know?

   no, because there is one extra chromosome in pair 21

2. Based on this karyotype chart, would you say the person has a genetic disorder? If so, what is the disorder?

   yes, Down's syndrome

Extension

Research the use of genetic markers, or small segments of DNA, to diagnose genetic diseases. Record your findings in the space provided.

Name _____ Class _____ Date _____
Chapter 31   Applied Genetics   Laboratory 31-1

Figure 1:   Human Chromosomes

Biology   Copyright © by Globe Book Company

Name _____ Class _____ Date _____

Chapter 31   Applied Genetics   Laboratory 31-1

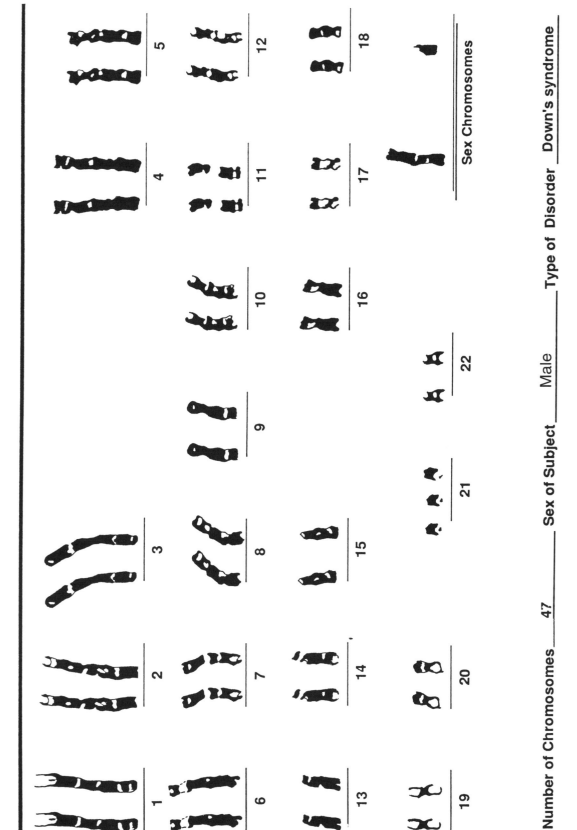

Number of Chromosomes __47__   Sex of Subject __Male__   Type of Disorder __Down's syndrome__

Figure 2:   Human Karyotype Chart

Biology   Copyright © by Globe Book Company   169

Name _____ Class _____ Date _____

Chapter 31   Applied Genetics

Laboratory 31-2 How can you make a model of controlled animal breeding?

Background Information

The process of breeding animals with certain desirable traits to produce offspring that consistently have those traits is called controlled breeding. For example, cat fanciers carefully breed show cats to meet specific physical standards that are characteristic of that particular breed of cat. Cats called Manx cats are the only cats born without tails. Sometimes Manx kittens are born with short tails, but a true Manx cat has no tail at all. Because inheritance of the trait for taillessness is determined by incomplete dominance, L represents the gene for long tail and L' the gene for no tail. Pure Manx cats have the genotype L'L'.

Skills: modeling, analyzing

Objective

In this laboratory, you will
- use a model to show how controlled breeding can be used to produce animals with certain traits.

Prelab Preparation

1. Review the term "incomplete dominance."

   Incomplete dominance is a blending of the traits carried by two different genes.

   _____

2. Review the use of Punnett squares to predict genetic probabilities.

Materials

construction paper
scissors
coin

Procedure

1. Use scissors to cut out eight squares of construction paper. **Caution: Be careful when using scissors.**

2. Mark an L on four of the squares and an L' on the remaining four squares.

3. Randomly place all eight squares face down on your work surface.

4. Choose two squares at random. The letters represent the genotype of a kitten in the F_1 generation. Record the genotype in Data Table 1.

5. Toss a coin to determine the sex of the kitten (heads = female, tails = male). Record the sex of the kitten in Data Table 1.

6. Repeat steps 3, 4, and 5 until you have recorded the genotype and sex of all four possible offspring in the F_1 generation.

Biology   Copyright ©   by Globe Book Company   171

7. Examine Data Table 1 to find a male and a female kitten with the genotype L'L'.

8. If you do not have both a male and female with the genotype L'L' in the F_1 generation, repeat steps 3, 4, and 5.

9. Record the number of generations needed to produce a male and a female cat with the desired genotype L'L'.

Observations and Data

Data Table 1: Offspring

F_1 Kittens	Genotype	Sex
1	Answers will vary.	
2		
3		
4		

Analysis and Conclusions

1. How many generations were needed to produce male and female kittens with the genotype L'L'?

   Answers will vary.

2. Why was this laboratory exercise an example of controlled breeding?

   because the model represented the number of generations of cats that a breeder

   would have to raise in order to produce one pair of cats with the desired trait

Extension

Animal breeders have developed a type of sheep that produces more wool than other sheep. The genotype of these sheep is FFmm. Construct a model to determine how many generations of selective breeding would be necessary to produce a pair of sheep with the genotype FFmm. Record your results in a data table.

Name _____ Class _____ Date _____

Chapter 32   Theories of Evolution

Laboratory 32-1 How do individuals of the same species vary?

Background Information

Variations are differences in traits that occur among members of the same species. Most variations are neutral; that is, they do not affect the individual in any way. Other variations are helpful; they help the individual survive in its environment. Helpful variations are called adaptations.

Skills: observing, inferring, hypothesizing

Objective

In this laboratory, you will
- observe some variations among members of different species.

Prelab Preparation

1. Review Section 32-1 Evolution.

2. Review the technique for drawing a line graph.

Materials

metric ruler	conifer twig	graph paper
paper	pencil	10 pinto beans

Procedure

Part A: Variations in Length of Conifer Needles

1. Remove 10 needles at random from a conifer twig.

2. Measure the length of each needle in millimeters.

3. Record the length of each needle in Data Table 1 by making a slash mark (/) in the "Number" column next to the correct measurement.

4. Add up the number of slash marks for each measurement and record these totals in Data Table 1.

5. Combine your results with the other students in your class. Record the class totals in Data Table 1.

6. Display the class data in a line graph. Record the number of needles along the vertical axis and the length in millimeters along the horizontal axis.

Part B: Variations in Length of Pinto Beans

1. Select 10 pinto beans at random.

2. Measure the length of each pinto bean in millimeters.

3. Record the length of each pinto bean in Data Table 2 by making a slash mark (/) in the "Number" column next to the correct measurement.

4. Add up the number of slash marks for each measurement and record these totals in Data Table 2.

Biology   Copyright ©   by Globe Book Company

5. Combine your results with the other students in your class. Record the class totals in Data Table 2.

6. Display the class data in a line graph. Record the number of pinto beans along the vertical axis and the length in millimeters along the horizontal axis.

Observations and Data

Data Table 1: Length of Conifer Needles

Length (mm)	Number	Total	Class Total
8		Answers will vary.	
9			
10			
11			
12			
13			
14			
15			
16			
17			
18			
19			
20			
21			

Name _____ Class _____ Date _____

Chapter 32 Theories of Evolution
Laboratory 32-1

Data Table 2: Length of Pinto Beans

Length (mm)	Number	Total	Class Total
5		Answers will vary.	
6			
7			
8			
9			
10			
11			
12			
13			
14			
15			
16			
17			
18			
19			

Analysis and Conclusions

1. a. Refer to Data Table 1. What is the average length of the conifer needles you measured?

   Answers will vary.

   b. What is the average length of all the conifer needles in the class?

   Answers will vary.

Biology   Copyright ©   by Globe Book Company   175

2. a. Refer to Data Table 2. What is the average length of the pinto beans you measured?

    Answers will vary.

    b. What is the average length of all the pinto beans in the class?

    Answers will vary.

3. Describe the shape of the line graph for a. the conifer needles and b. the pinto beans.

    Both graphs are bell shaped.

4. What does the shape of the graphs indicate about the lengths of the conifer needs and the pinto beans?

    The lengths vary, but most are near the average length, with a few individuals at each end of the graph.

5. Describe the variations you observed in the lengths of (a) the conifer needles and (b) the pinto beans.

    Answers will vary depending on student data. Students should include a range of lengths for both the conifer needles and the pinto beans.

Extension

What are the variations in the width of the hand among students in your class? Have each student stretch the fingers of his or her right hand flat on a desktop. Measure the distance from the tip of the thumb to the the tip of the finger. Record the measurements in a data table. Calculate the average width and display your data on a line graph.

Name _____ Class _____ Date _____

Chapter 32 — Theories of Evolution

Laboratory 32-2 What is natural selection?

Background Information
Natural selection is the process by which organisms that are well–adapted to their environment have a better chance of surviving and reproducing than do organisms that are poorly adapted. Natural selection is often referred to as "survival of the fit." Well–adapted organisms that survive to reproduce pass on their traits to their offspring.

Skills: modeling, inferring

Objective
In this laboratory, you will
- use a model to demonstrate natural selection in giraffes.

Prelab Preparation

1. Review Section 32–1 Evolution and Section 32–2 Evolution and Genetics.

2. Review the following procedures for your model:

   a. Assume that a population of giraffes is living in an area with tall trees. Long–necked giraffes can reach and eat the leaves on these trees; they are strong and healthy. Short–necked giraffes cannot reach the leaves and do not get enough to eat. As a result, they are weak and are easily caught by predators.

   b. Two cards from the deck of gene cards together represent the genotype of a giraffe. The letter "L" represents the dominant gene for a long neck.

   c. A predator card marked with a letter "P" means that a predator is after your giraffe. If you choose a "P" card and you have a long–necked giraffe, your giraffe escapes and lives to reproduce.

   d. If you choose a "P" card and you have a short–necked giraffe, your giraffe is killed by the predator. Its genes are removed from the populations.

   e. If you choose a blank predator card, both long– and short–necked giraffes live to reproduce.

Materials
60 gene cards
30 predator cards

Note to teacher: Use index cards to prepare two decks of cards. Write a captial "L" on 30 gene cards and a lower case "l" on the remaining 30 gene cards. Write the letter "P" on 20 predator cards and leave the remaining 10 cards blank.

Procedure

1. Choose two cards at random from the deck of gene cards.

2. Report the length of your giraffe's neck to your teacher, who will record the class data in a data table.

   Note to teacher: You may wish to assign one student to record class data.

3. Choose a predator card at random.

4. Report whether you chose a "P" card or a blank card. If you chose a "P" card and you had a short–necked giraffe, remove your two gene cards from the deck.

Biology   Copyright © by Globe Book Company   177

5. If you chose a "P" card and you had a long–necked giraffe, or if you chose a blank card, return your gene cards to the deck.

6. Return your predator card to the deck.

7. Count the number of gene pairs remaining in the deck after the first generation.

Note to teacher: Collect the gene cards after each generation. Prepare a new set of cards that has the same ratio of long necks to short necks as in the previous generation.

8. Repeat steps 1 to 7 four more times for a total of five generations of giraffes.

9. Complete Data Table 1 by recording the class data.

Observations and Data

Data Table 1: Generations of Giraffes

	First		Second		Third		Fourth		Fifth	
	long neck	short neck	long neck	short neck	long neck	short neck	long neck	short neck	long neck	short neck
Number of giraffes										
Number of giraffes that live to reproduce					Answers will vary.					
Number of gene pairs that remain										

Analysis and Conclusions

1. What happened to the ratio of long-necked giraffes to short-necked giraffes between the first and second generations? between the first and third generations? between the first and fifth generations? Use your data to support your answers.

   The proportion of long–necked giraffes increased after each generation.

2. What happened to the ratio of "L" genes to "l" genes in the population over five generations?

   The ratio of "L" genes to "l" genes increased.

3. How do you think this model demonstrates natural selection?

   Long–necked giraffes, which are well–adapted to their environment, survived to reproduce, passing on their genes to their offspring.

Extension

Choose a different population of animals, with different traits, living in a different environment. Prepare gene cards with a capital letter for the dominant trait and a lower case letter for the recessive trait. Repeat this laboratory for this population of animals. Record your data in a data table.

Name _____ Class _____ Date _____

Chapter 33   Evidence for Evolution

Laboratory 33-1 What is the half–life of a radioactive element?

Background Information

Carbon–14 is a radioactive element that breaks down, or decays, into nitrogen. The time it takes for one–half of the carbon–14 to break down into nitrogen is called its half–life. The carbon–14 in an organism begins to decay into nitrogen when the organism dies. For this reason, the half–life of carbon–14 can be used to find the absolute age of a fossil by comparing the amount of carbon–14 with the amount of nitrogen in the fossil.

Skills: modeling, hypothesizing

Objective

In this laboratory, you will
- use a model to demonstrate the half–life of an element.

Prelab Preparation

Review Section 33–1 Fossil Evidence.

Materials

metric ruler   pencil
scissors   paper

Procedure

1. Draw a square 10 cm by 10 cm on a piece of paper. Use scissors to cut out the square. **Caution: Be careful when using scissors.** This square represents the carbon–14 in a fossil.

2. Cut the square in half. One–half of the square represents the amount of carbon–14 that has not decayed. The other half represents nitrogen. Discard the "nitrogen" half.

3. Refer to Data Table 1. Make a slash mark (/) in the "Number" column each time you cut the paper in half.

4. Continue to cut the paper in half, discarding the "nitrogen" half after each cut, until the paper is too small to cut any further. Remember to make a slash mark in Data Table 1 after each cut.

Observations and Data

Data Table 1:   Half-Life of Carbon-14

Number of Cuts	Half-Life of Carbon-14	Number of Years/Cut	Total Years
Answers will vary.	5700 years	5700 years	5700 x number of cuts

Biology   Copyright ©   by Globe Book Company   179

1. How many cuts were you able to make in your paper sample?

    Answers will vary.

2. What is the absolute age of your fossil?

    Answers will vary. Students should multiply the number of cuts by 5700, the half-life of carbon-14.

Analysis and Conclusions

1. What did each cut represent in this model?

    one half-life

2. Which of the following fossils do you think is older: a fossil containing 25% carbon–14 and 75% nitrogen or a fossil containing 50% carbon–14 and 50% nitrogen? How do you know?

    The fossil containing 25% carbon-14 is older because more of the carbon-14 has decayed into nitrogen.

3. Do you think carbon–14 dating could be used to find the age of a dinosaur fossil? Why or why not?

    No, because the dinosaurs lived over 65 million years ago, and the half-life of carbon-14 is not long enough.

Extension

Use reference materials to find out the half–life of uranium. Then repeat this laboratory, using uranium instead of carbon-14. Use reference materials to research other radioactive elements that are used in radioactive dating. Record your findings in a written report.

Name _____ Class _____ Date _____

Chapter 33 — Evidence for Evolution

Laboratory 33-2 How do living things provide evidence for evolution?

Background Information

Body parts in different organisms that have the same basic structure are called homologous structures. By comparing homologous structures, biologists can determine how organisms are related. The presence of homologous structures suggests that organisms evolved from a common ancestor.

Skills: comparing and contrasting, analyzing, hypothesizing

Objective

In this laboratory, you will
- compare and contrast homologous structures in living things.

Prelab Preparation

1. Review Section 33–3 Evidence from Living Things.

2. Compare the meaning of the terms "homologous" and "analogous."

    "Homologous" means "having the same relative structure." "Analogous" means "having a similar function

    but different structure and origin."

Materials

paper   pencil

Procedure

1. Refer to Figure 1. Carefully examine the relative size, shape, number, and position of the bones in the body parts shown.

2. In Data Table 2, identify the body part shown for each organism listed. Then describe the function of each body part.

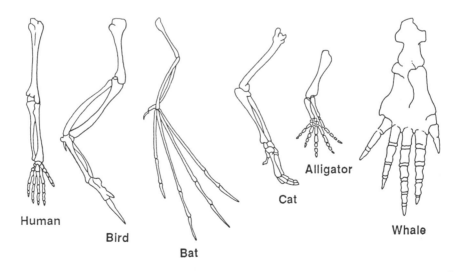

Figure 1: Homologous Structures

Biology   Copyright © by Globe Book Company   181

3. Refer to Figure 2. Compare the structure of a bird wing and a butterfly wing shown in the drawing.

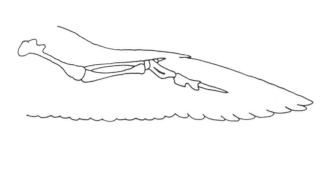

Bird wing   Butterfly wing

Figure 2: Analogous Structures

Observations and Data

Data Table 1: Homologous Structures

Organism	Body Part	Function
Human	arm	holding, grasping
Whale	flipper	swimming
Cat	foreleg	walking
Bat	wing	flying
Bird	wing	flying
Alligator	foreleg	walking

1. How are the body parts shown in Figure 1 alike? How are they different?
   Answers will vary. Accept all logical responses.

2. Do the body parts in Figure 2 have similar or different structures?
   different

3. Do the body parts in Figure 2 have similar or different functions?
   similar

Name _____ Class _____ Date _____

Chapter 33 Evidence for Evolution

Laboratory 33-2

Analysis and Conclusions

1. Why are the structures shown in Figure 1 homologous structures?
   because they have the same basic structure

2. Why are the structures shown in Figure 2 analogous structures?
   because they are not similar in structure, but perform the same basic function

3. How do you think the arm of a human and the flipper of a whale show evidence of an evolutionary relationship?
   Each has evolved to help the organism survive in its environment; the similarity in structure suggests evolution from a common ancestor.

4. Do the wings of a bird and a butterfly suggest an evolutionary relationship? How do you know?
   No, because they have no structural similarities although they evolved to perform a similar function.

Extension

Use reference materials to research the evolutionary relationship between dinosaurs and birds. What evidence suggests that birds may have evolved from dinosaurs? Record your findings in a written report. Use the space provided on the following page.

Biology   Copyright © by Globe Book Company   **183**

Notes

Name_____ Class_____ Date_____

Chapter 34 Human Change Through Time

Laboratory 34-1 How are humans similar to and different from other primates?

Background Information
Humans are classified as primates. Humans and other primates have many characteristics in common, which led Charles Darwin to hypothesize that they had evolved from a common ancestor. Recently, biochemists have found similarities in the sequence of amino acids that make up certain proteins in different primates. By comparing these sequences, scientists can infer that some primates are more closely related to humans than are other primates.

Skills: comparing and contrasting, inferring

Objective
In this laboratory, you will
- infer how closely humans are related to other primates.

Prelab Preparation
1. Review Section 34–1 Classification of Humans.
2. What are amino acids?

   the building blocks of proteins

Procedure
1. Examine Data Table 1, which lists the sequence of amino acid molecules in the blood protein hemoglobin for humans and other primates.

2. Note the position of each amino acid. Count the number of similarities in the position of each amino acid in human protein compared with the position in the proteins of the other primates. Record the number of similar amino acid positions in Data Table 2.

3. Count the number of differences in the position of each amino acid in human protein compared with the position in the proteins of the other primates. Record the number of different amino acid positions in Data Table 2.

Observations and Data

Data Table 1: Amino Acid Sequences in Hemoglobin

Human	Chimpanzee	Gorilla	Baboon	Lemur
SER	SER	SER	ASN	ALA
THR	THR	THR	THR	THR
ALA	ALA	ALA	THR	SER
GLY	GLY	GLY	GLY	GLY
ASP	ASP	ASP	ASP	GLU
GLU	GLU	GLU	GLU	LYS
VAL	VAL	VAL	VAL	VAL
GLU	GLU	GLU	ASP	GLU
ASP	ASP	ASP	ASP	ASP
THR	THR	THR	SER	SER
PRO	PRO	PRO	PRO	PRO
GLY	GLY	GLY	GLY	GLY
GLY	GLY	GLY	GLY	SER
ALA	ALA	ALA	ASN	HIS
ASN	ASN	ASN	ASN	ASN
ALA	ALA	ALA	ALA	ALA
THR	THR	THR	GLN	GLN
ARG	ARG	LYS	LYS	LYS
HIS	HIS	HIS	HIS	LEU

Biology   Copyright © by Globe Book Company

Data Table 2: Similarities and Differences in Amino Acid Positions

Organisms	Number of Similarities	Number of Differences
Human and chimpanzee	19	0
Human and gorilla	18	1
Human and baboon	12	7
Human and lemur	9	10

Analysis and Conclusions

1. Which of the primates listed in Data Table 1 had the most similarities with humans? Which had the fewest?

   chimpanzee; lemur

2. Which primate is most closely related to humans? How do you know?

   chimpanzee, because it has the greatest number of similarities in amino acid positions

3. Which primate is least closely related to humans? How do you know?

   lemur, because it has the fewest similarities and most differences in amino acid positions

Extension

Use reference materials to compare the similarities and differences in amino acid positions between humans and other orders of mammals. Record your results in a data table.

Name _____ Class _____ Date _____

Chapter 35 Ecosystems

Laboratory 35-1 What are food chains?

Background Information
The pathway that energy, in the form of food, takes through an ecosystem is called a food chain. A food chain shows how energy moves from producers to first–, second–, and third–level consumers.

Skills: comparing, sequencing, inferring

Objective
In this laboratory, you will
- trace the flow of energy from producers to consumers in food chains.

Prelab Preparation
Review Section 35–2 Energy and the Environment.

Materials
paper   pencil

Procedure
1. Refer to Data Table 1, which lists some organisms in a forest–pond ecosystem.
2. For each animal listed down the left side of the table, indicate what organisms the animal eats by making a check mark (√) in the column under the correct heading.
3. Complete Data Table 1 for each animal listed.

Observations and Data

Data Table 1: Energy in an Ecosystem

Organisms in a Forest-Pond Ecosystem	What Forest-Pond Organisms Eat												
	Cricket	Earthworm	Hawk	Insects	Land snail	Mouse	Owl	Plants	Rabbit	Raccoon	Robin	Snake	Sparrow
Cricket								√					
Earthworm								√					
Fox		√				√			√		√		√
Hawk						√			√			√	√
Insects								√					
Land snail								√					
Mouse	√			√	√			√					
Owl						√			√			√	
Rabbit								√					
Raccoon						√		√					
Robin	√	√											
Snake		√				√			√		√		
Sparrow								√					

Analysis and Conclusions

1. Which of the organisms in Data Table 1 are producers?
   plants

2. Which organisms are herbivores?
   cricket, earthworm, insects, land snail, rabbit, sparrow

3. Which organisms are carnivores?
   snake, fox, hawk, owl, robin

4. Which organisms are omnivores?
   mouse, racoon

5. Which organisms are first–level consumers?
   cricket, earth worm, insects, land snail, rabbit, sparrow

6. Which organisms are second–level consumers?
   mouse, robin, snake, hawk

7. Which organisms are third–level consumers?
   snake, fox, hawk, owl, racoon

Extension

Use Data Table 1 to construct several food chains. Combine your food chains into a food web. Draw an energy pyramid for your food web.

Name _____ Class _____ Date _____

Chapter 35   Ecosystems

Laboratory 35-2   What is the relationship between a predator and its prey?

Background Information

All the relationships among organisms in an ecosystem help to keep the ecosystem in balance. One of the most important kinds of relationships in an ecosystem is the predator–prey relationship. A predator is an animal that kills and eats other animals. The prey is the animal that is killed by the predator.

Skills: observing, predicting, inferring

Objective

In this laboratory, you will
- observe a predator–prey relationship.

Prelab Preparation

1. Review Section 35-3 Relationships in an Ecosystem.

2. Review the procedure for using a compound microscope.

3. Review the procedure for making a wet–mound slide.

Materials

compound microscope   depression slide
cover slip   3 droppers
paramecium, *Chlorella*, and *Didinium* cultures

Procedure

1. Place one drop each of paramecium, Chlorella, and Didinium cultures onto a clean, dry depression slide. Use a clean dropper for each culture. Place a cover slip over the depression.

2. Place the slide on the stage of a compound microscope. Observe the slide under low power. Then switch to high power.

3. Count the number of paramecia, *Chlorella*, and *Didinium* in the field of view. Record these numbers under the heading "0 minutes" in Data Table 1.

4. Predict if the number of paramecia, *Chlorella*, and *Didinium* will increase or decrease after 5 minutes, 10 minutes, and 15 minutes. Record your predictions in Data Table 1. Place a plus sign (+) in the correct column if you think the number will increase; place a minus sign (–) in the column if you think the number will decrease.

5. After 5 minutes, observe the slide and count the number of each organism in the field of view. Record these numbers in Data Table 1.

6. Repeat step 5 after 10 minutes and again after 15 minutes.

Biology   Copyright ©   by Globe Book Company

Observations and Data

Data Table 1: Number of Organisms

Organisms	0 minutes	5 minutes		10 minutes		15 minutes	
	Actual	Predicted	Actual	Predicted	Actual	Predicted	Actual
Chlorella				Answers			
Paramecia				will			
Didinium				vary.			

Were your predictions correct?

Answers will vary.

Analysis and Conclusions

1. Which organisms in this laboratory was the predator?

    Didinium

2. Which organism was the prey?

    paramecium

3. Which of the three organisms is a producer?

    Chlorella

4. Which of the organisms are consumers?

    paramecium and _Didinium_

5. Draw a food chain describing the relationships you observed in this laboratory.

    Chlorella → **paramecium** → *Didinium*

Extension

How does time affect the relationships among populations of *Chlorella*, paramecia, and *Didinium*? Design an experiment to compare the numbers of *Chlorella*, paramecia, and *Didinium* over a period of three days. Record your observations in a data table.

Name_____ Class _____ Date _____

Chapter 36 — Biomes

Laboratory 36-1 How are organisms adapted to survive in different biomes?

Background Information

A biome is a large region of the earth where characteristic kinds of organisms live. What makes an organism better suited to life in one biome than in another? Organisms have adaptations that help them to survive in a particular biome. Adaptations are helpful variations found among the members of a population.

Skills: inferring, classifying

Objective

In this laboratory, you will
- match several organisms and their adaptations with a specific biome.

Prelab Preparation

Review Section 36–1 Land Biomes.

Materials

paper
pencil

Procedure

1. Refer to Data Table 1. Study the lists of plants, animals, and adaptations.

2. Use the information in Data Table 1 to complete Data Table 2. List three plants and three animals for each biome. Then list one adaptation for each organism that helps the organism survive in that biome.

Observations and Data

Data Table 1: Plants, Animals, and Adaptations

Common plants (some may be found in more than one biome)

hickory trees	grasses	vines
palms	mosses	small bushes
oak trees	beech trees	ferns
lichens	wildflowers	creosote
spruce	pine trees	maple trees

Common animals (some may be found in more than one biome)

caribou	deer	antelope
snakes	lizards	monkeys
insects	lynx	musk oxen
squirrels	prairie dogs	armadillos
moose	polar bears	anteaters
rodents	black bears	

Adaptations (some may be common to more than one organism)

nocturnal	spines for leaves
needle–like leaves	grazing animals
burrowers	shallow roots
grow low to the ground	require short growing season
require much moisture	cone–bearing
require little moisture	lose leaves to protect against
coat changes with seasons	loss of water
short life cycle	prey on small mammals
leaves all year	can survive dry periods

Name _____ Class _____ Date _____

Chapter 36 Biomes   Laboratory 36-1

Data Table 2: Adaptations of Plants and Animals

Biome	Common Plants	Adaptations	Common Animals	Adaptations
Tundra	1. 2. 3.	1. 2. 3.	1. 2. 3.	1. 2. 3.
Taiga	1. 2. 3.	1. 2. 3.	1. 2. 3.	1. 2. 3.
"Spruce–moose" belt	1. 2. 3.	1. 2. 3.	1. 2. 3.	1. 2. 3.
Southern pine forest	1. 2. 3.	1. 2. 3.	1. 2. 3.	1. 2. 3.
Deciduous forest	1. 2. 3.	1. 2. 3.	1. 2. 3.	1. 2. 3.
Tropical rain forest	1. 2. 3.	1. 2. 3.	1. 2. 3.	1. 2. 3.
Grassland	1. 2. 3.	1. 2. 3.	1. 2. 3.	1. 2. 3.
Desert	1. 2. 3.	1. 2. 3.	1. 2. 3.	1. 2. 3.

Biology   Copyright © by Globe Book Company

Analysis and Conclusions

1. How are biomes identified?

   Biomes are identified by their dominant plant life.

2. How does the plant life in a biome affect the kinds of animals found in that biome?

   The kinds of animals found in a biome are determined by the kinds of plants that grow in that biome.

3. What determines the kinds of biomes found in a particular area?

   climate, or average yearly rainfall and temperature

Extension

Use reference materials to research the greenhouse effect, or global warming. How might global warming affect the earth's biomes? Record your findings in a written report.

Name _____ Class _____ Date _____

Chapter 37 — Conservation

Laboratory 37-1 How do pollutants in water affect the germination of seeds?

Background Information

Water is one of the most important natural resources on earth. Crop plants need a regular supply of clean water to grow. Water pollution reduces the amount of clean water available. Water pollution can be caused by phosphates in detergents, fertilizers and pesticides, and toxic wastes from industry.

Skills: observing, comparing, measuring

Objective

In this laboratory, you will
- observe the effects of different pollutants on the germination of lima bean seeds.

Prelab Preparation

1. Review Section 37–2 Three Renewable Resources.

2. Review how to calculate a percentage.

Note to teacher: Prepare the solutions and soak the lima beans for 24 hours before students begin this laboratory. Assign each group of students a different pollutant and have them compare and share their data.

Materials

lima beans	masking tape	water
household ammonia	detergent	vinegar
sulfuric acid	paper towels	marking pencil
plastic sandwich bags		

Procedure

1. Place 10 lima beans that have been soaked in water on a paper towel, as shown in Figure 1.

2. Fold the paper towel over the lima beans.

3. Use a dropper to moisten the towel with water.

4. Slide the folded paper towel and lima beans into a plastic sandwich bag.

5. Seal the bag with a piece of masking tape. Label the tape with the word "water."

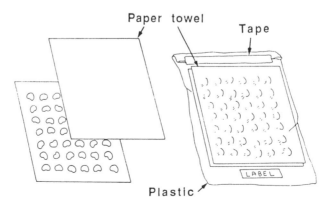

Figure 1: Germinating Lima Beans

Biology — Copyright © by Globe Book Company — 195

6. Repeat steps 1 to 5 with 10 lima beans that have been soaked in one of the four pollutants listed in Data Table 1. Label the tape with the name of the pollutant in which the lima beans were soaked.

7. Put the bags in a warm, dark place where they will not be disturbed for two days.

8. After two days, observe the lima beans for signs of germination.

9. Count the number of lima beans that germinated in each bag. Record these numbers in Data Table 1.

10. Complete Data Table 1 for the remaining pollutants by sharing data with other groups in your class.

11. Calculate the percentage of lima beans that germinated.

Observations and Data

Data Table 1: Germination of Lima Beans

Lima Beans Soaked in	Number Used	Number Germinated	Percentage Germinated
Water	10	Answers will vary.	
Ammonia	10		
Sulfuric Acid	10		
Vinegar	10		
Detergent	10		

1. Which sample of lima beans showed the highest percentage of germination?
   lima beans soaked in water

2. Which sample showed the lowest percentage of germination?
   Answers will vary.

3. Arrange the lima bean samples in sequence, from highest percentage of germination to lowest.
   Answers will vary.

Name _____ Class _____ Date _____
Chapter 37   Conservation   Laboratory 37-1

Analysis and Conclusions

1. How did the pollutants used in this laboratory affect the germination of lima beans?
   The percentage of germinated beans soaked in pollutants was lower than the percentage of germinated

   beans soaked in water.

2. What other household chemicals might have the same effect on the germination of lima beans?
   Answers will vary. Answers may include bleach, insecticides, drain cleaner, shampoo, etc.

Extension

Could the amount of pollutant used to make the solutions in which the lima beans were soaked have an effect on the results of this laboratory? Design an experiment to answer this question. Record your results in a data table. Use the space provided on the following page to describe your experiment.

Biology   Copyright ©   by Globe Book Company   197

Notes

Name _____ Class _____ Date _____

Chapter 37   Conservation

Laboratory 37-2 What kinds of pollutants are in the air?

Background Information

Air pollution is a mixture of gases and particulates. Most particulates result from burning, windblown dirt, dust, or pollen, worn machinery, and the brake linings and tires of automobiles. These particulates can damage the lungs and cause respiratory diseases.

Skills: classifying, inferring, analyzing

Objective

In this laboratory, you will
- collect and observe particles from the air around your school.

Prelab Preparation

1. Review Section 37–2 Three Renewable Resources.

2. Review the procedure for using a compound microscope.

Note to teacher: You may wish to have different groups of students collect data from different locations in and around the school building. Students can then compare and share data.

Materials

compound microscope
3 microscope slides
clear tape
petroleum jelly

graph paper
marking pencil
scissors

Procedure

1. Trace the outline of a microscope slide on a piece of graph paper. Repeat so that you have a total of three outlines.

2. Use scissors to cut out each of the three outlines. **Caution: Be careful when using scissors.**

3. Tape one outline to each microscope slide so that the graph paper grid is visible through the slide. Label the slides 1, 2, and 3 on the back of the graph paper.

4. Cover the surface of each slide with a thin coating of petroleum jelly.

5. Leave each slide in a different location where it will not be disturbed overnight.

6. Refer to Data Table 1. Record the location for each of your slides.

7. Predict which of the three locations will yield the most particles after 24 hours and which location will yield the fewest. Record your predictions in the space provided.

8. After 24 hours, collect your slides. Handle the slides carefully so as not to disturb the petroleum jelly.

9. Place a slide on the stage of a compound microscope and observe it under low power. Be careful not to let the lens touch the petroleum jelly.

10. Observe the particles on the slide and classify them as dirt, dust, pollen, fibers, or other. Record your obeservations in Data Table 1.

Biology   Copyright ©   by Globe Book Company   199

11. Count the number of particles in 10 squares of the graph paper. If a particle or fiber extends into two squares, count it as two particles.

12. Have your partner count the number of particles in 10 squares. Find the average of the two numbers, yours and your partner's. Record the average number of particles in Data Table 1.

13. Repeat steps 9 to 12 for the other two slides. Complete Data Table 1.

Observations and Data

Predicted order of locations:

Most particles **Answers**

will

Fewest particles **vary.**

Data Table 1: Particles in the Air

Location	Average Number	Kind of Particles	Description of Particles
1.			
2.			
3.			

1. Were your predictions correct?

   Answers will vary.

2. What kind of particle was most common on all three of your slides?

   Answers will vary.

3. Compare your data with other groups in your class. Which locations yielded the greatest number of particles? Which location yielded the least number of particles?

   Answers will vary.

4. What kind of particle was found most frequently on all the slides in your class?

   Answers will vary.

Name _____ Class _____ Date _____

Chapter 37 Conservation

Laboratory 37-2

Analysis and Conclusions

1. How can you explain the differences in the numbers and kinds of particles found in different locations?
   Answers will vary. Accept all logical responses.

2. Based on your observations, are there any locations in or around your school that might present a health problem?
   Answers will vary. Accept all logical responses.

Extension

How would the results of this laboratory be different if you placed your slides in different locations? Repeat this laboratory, placing slides in various locations, such as a busy intersection, a park, a room in your house, and so forth. Predict which location will yield the most particles. Record your results in a data table. Use the space provided on the following page to describe your results.

Biology   Copyright © by Globe Book Company   **201**

Notes